Telepresence: Actual and Virtual

Telepresence: Actual and Virtual

Promises and Perils of Converging New Realities

Thomas B. Sheridan

CRC Press
Taylor & Francis Group
Boca Raton London New York

CRC Press is an imprint of the
Taylor & Francis Group, an **informa** business

First edition published 2023
by CRC Press
6000 Broken Sound Parkway NW, Suite 300, Boca Raton, FL 33487-2742

and by CRC Press
4 Park Square, Milton Park, Abingdon, Oxon, OX14 4RN

CRC Press is an imprint of Taylor & Francis Group, LLC

ISBN: 978-1-032-27943-5 (hbk)
ISBN: 978-1-032-28628-0 (pbk)
ISBN: 978-1-003-29775-8 (ebk)

DOI: 10.1201/9781003297758

Typeset in Times
by Deanta Global Publishing Services, Chennai, India

Contents

Preface

There are three reasons why I chose to write this book.

1. First, robotic technology, along with telecommunication technology, has now matured, enabling humans to operate devices that are physically remote from themselves. 5G *and* 6G internet communication will revolutionize the connectivity of our society and will allow for remote access to a wide range of services, democratizing accessibility. Also matured is the sensing technology that can accompany the remote robot, allowing sufficient visual, auditory, tactile, and kinesthetic sensation to be presented back to the human operator via specialized displays. Thus, the human experiences are present at the remote location, known as "telepresence". In a closely connected realm of technology application, computer graphic simulation software has developed to the point where human operators can experience a similar sense of presence in an environment that is fully computer generated and therefore totally virtual. This "virtual reality" capability allows humans to come together in virtual spaces, for them to visit and physically move through such virtual places anywhere in the world that are recorded by computers, such as museums or natural sites, or to control avatar beings in games and meetings. Finally, a technology called "augmented reality" is emerging, allowing the actual remote and the computer-generated virtual realities to be combined in the visual display. These technologies have much in common, as should become evident as the successive book chapters look at the history, philosophy, technology applications, and what I call "challenges" for the future. This book hopes to provide readers some appreciation of what these technologies are and what they portend. The reader will discover that both robotic telepresence and virtual reality are about *extending one's reach to interact with things and people*, and how that newfound capability has both benefits and costs, promises, and perils.
2. Second, what good can come of telepresence and virtual reality? While the above technologies have begun to find their way into the market, investment and use have thus far been sporadic and limited in application. However, recent advances in digital communication technology have heralded a huge increase in the so-called *internet-of-things*, through which devices in factories, homes, highways, or hospitals can be controlled remotely. The recent pandemic motivated the wide use of teleconferencing technology, like Zoom. Now the huge technical corporations, such as Meta (Facebook), Microsoft, Apple, Alphabet (Google), and Amazon are reporting major investments in what is being hyped as the *metaverse*. One certain investment will result in greater appeal and sophistication of computer-game playing, an activity heretofore of questionable social value. But games have future potential for bringing people together for education and community

involvement. Beyond game playing, I believe there will be a variety of new and promising applications of robotic telepresence devices, such as telemedicine. There have also been applications that not only replace undesirable human labor, but by being controlled remotely can avoid travel and the associated undesirable energy costs. Having had a small hand in the early development of the above technologies, I naturally look forward to their usefulness in many functions of modern society, particularly as a means to avoid the energy costs of physical travel. How might we shape the directions of investment and application of these technologies to do the most good?

3. Finally, however, I am genuinely fearful of unintended consequences. I am fearful of the ease by which anyone or any group might commit mischief or even criminal activity via anonymous remote control in the real world or control of an avatar in the virtual world. This extends from unwanted sexual groping to illegal spying to destruction and acts of war—both by remotely controlled robots and by activities in seductive virtual environments. The fact that military planners are increasingly attracted to this technology does not give comfort. I would hope that this book might alert responsible individuals to the nature, potential pitfalls, and significant costs of telepresence and virtual reality technologies. There is a need for public discussion, anticipation, and government involvement. I make no claim that this book is an up-to-date review of the technology details of telepresence and virtual reality. My principal purpose and hope is that this book, in laying out the history, the general nature of the technology, the great variety of potential applications, and what we have learned from psychological experiments and philosophical debates about "reality", can edify the public and encourage needed social and regulatory constraints.

Acknowledgments

I am grateful to Hari Das Nayar, Adrian Stoica, Edward Tunstel, Farokh Atashzar, and Woodrow Barfield for initial support and for commenting on early drafts of the book. I also want to thank my many graduate students over the years, who did the hard work in my own MIT lab, some of which is reported in the book. Finally, I acknowledge the love and patience of Nancy, who read through my turgid prose, and wondered why, at 92 years old, I needed to write a tenth book.

About the Author

Thomas B. Sheridan is Ford Professor Emeritus, formerly Professor of Engineering and Applied Psycholgy, in the Department of Mechanical Engineering and Department of Aeronautics and Astronautics at the Massachusetts Institute of Technology (MIT). He is a pioneer of robotics and remote control technology. In 1951, he earned his B.S. degree in Mechanical Engineering from Purdue University, an M.S. Eng. degree from the University of California, Los Angeles, in 1954, and an Sc.D. degree from MIT in 1959. He was awarded an honorary doctorate from Delft University of Technology in the Netherlands and is a member of the National Academy of Engineering. His research interests are in experimentation, modeling, and design of human–machine systems in air, highway, and rail transportation, space and undersea robotics, process control, arms control, telemedicine, and virtual reality. Together with his graduate students in the Human-Machine Systems Lab at MIT, Sheridan developed important concepts concerning human–robot interaction, particularly regarding supervisory control and telepresence.

Introduction

Human beings are social animals. We like to be together—to talk in spoken language and body language, to hug, to make love, and just hang out together. We also like to do things for ourselves, hands on—to raise children or crops, to garden, to cook, to engage in hobbies, to go to concerts and museums, and to visit the family. Part of doing things ourselves, especially in cooperation with other people, is to feel productive—doing for others, what psychologist Erich Fromm called "productive orientation". That's what life is all about.

However, there are compelling trends that work against this freedom for direct contact with other people and with the world: (1) One big factor is the demand for efficiency in doing jobs and making the best use of our time. A response to the first factor is the introduction of technology, in particular robots and automation. (2) We can't be everywhere, and we don't want to live exactly where the needed work is. A major response to the second factor is more technology—communication via computer, cell phone, and TV, and working from home. (3) A third factor is climate change. Traveling poses a cost, not only in time and money but also in energy use, which to some extent is climate compromising. (4) A fourth factor is health. While the recent pandemic may abate, it constrains us from engaging with other people as much as we would like.

The first factor, demand for efficiency and the use of automation technology, has been with us in one form or another since the start of the industrial revolution, several centuries. The incredible capabilities of electronic communication combined with computation are new. Inconveniences brought about by travel disincentives due to climate change will be with us for some time.

We are slowly facing up to the need for, and are witnessing the beginnings of, a major reorientation in our industry and lifestyle. We are working from home communicating with others via computer, so no need to go into the office. Across a wide range of tasks, for convenience and safety, robots are replacing people, and being controlled by those people from a distance at control stations physically separated from the action. Examples are on production lines, warehousing (think Amazon), telemedicine, use of zoom communication for meetings, and control of myriad devices over the "internet of things". MOOCS (massive open online computer systems) are finding increasing use in education. Robots have essentially replaced human astronauts. Mining, agriculture, construction, and even warfare are gradually seeing an invasion of special-purpose robots manned at a distance.

The robots need to be supervised by humans. No matter how smart the artificial intelligence of the robots, and in spite of all the popular press worries about robots taking over, at some level they depend on human supervision. What that requires is the human knowing what is going on with the robot, the particular task, and the environment of the task. In other words, the human–robot supervisor needs to sense all

DOI: 10.1201/9781003297758-1

the information that would be available were he or she present at the site of the robot and the task. This is called *telepresence*, the robot operator having the experience of "feeling like he is there" though he really, physically is not "there".

The technology and the problems involved in making telepresence happen are what this book is about. As with all technologies, telepresence technology has its benefits, and these benefits are promoted by the developers of the technology. Most of the benefits are obvious. But telepresence technology also has its costs and perils, and these are not so obvious. These will be reviewed, as best this author can anticipate. Definitions of relevant telepresence terms are stated below.

But before defining telepresence terms, it is important to consider a closely allied field of technology that has grown dramatically in recent years, what is called *virtual reality* (VR). This is computer technology, combined with means to display to the human user the visual, auditory, and haptic (touch and motion) senses within a fully computer-generated environment with which the human is interacting. The book will make clear how robotic telepresence for an actual physical task and virtual reality, where the task and its environment are artificially generated by computers, *have much in common*, primarily with respect to the nature of the interaction with the human user. The term *telepresence* aptly applies to virtual reality because *tele* (meaning at a distance) can easily refer to the "distance" from the real world experienced by the VR participant.

Major tech firms like Meta (Facebook), Microsoft, and Alphabet (Google) are currently gearing up for what they anticipate will be a major market for VR (New York Times, December 30, 2021), anticipating their use in education, tourism, sales, social interactions, and entertainment.

When virtual (computer-generated) images are superposed on the real physical environment with which the operator is interacting, then it is called *augmented reality* (AR) or mixed reality. Objects in the real world are enhanced by computer-generated perceptual information, which can be across multiple sensory modalities, including visual, auditory, haptic (touch and motion), somatosensory (e.g., warmth, pain, taste), and olfactory modalities. A particularly challenging problem with augmented reality is to make the digital imagery correspond spatially (vertically, horizontally, size, color, texture, motion) to the real-world image, since a discrepancy is easily noticeable and the effect of a combined image is then compromised.

I am neither a proponent nor an opponent of telepresence technology. My purpose in writing this book is to call attention to particular technological changes that will continue to expand and change our lives for better and worse. The book will review both the better and the worse. I am taking for granted some of the advantages that I see emerging. I am apprehensive of some of the negative effects and the unanticipated consequences. I will do my best to help anticipate them.

SOME DEFINITIONS

Telepresence refers to a set of technologies that allow persons to feel as if they were present, to give the appearance of being present, or to have an effect, via telerobotics, at a place other than their true location (Wikipedia). *Telepresence* uses technology that enables a person to perform actions in a distant or virtual location as if physically present in that location (Webster).

Telexistence, a term used by Japanese researchers, has the same meaning as telepresence.

Teleoperation is the control of a device or machine remotely (Oxford).

Telerobotics is the area of robotics concerned with the control of semi-autonomous robots from a distance. A **telerobot** is such a robot.

Telerobotics can be said to combine two major subfields: teleoperation and telepresence.

As in telephones, telegrams, and television, the "tele" prefix is from Greek *tēle*, meaning "far off" or "at a distance".

Computerized simulation, or just **simulation**, as the term is used in this book, refers to the capability of a digital computer to utilize stored or real-time data, make calculations, generate computer graphics, and render to human visual, auditory, tactile, or kinesthetic senses, the images or patterns that convey meaning in the given context.

Computer graphics and computerized simulation have a history as long as that of computers. Well before what we call virtual reality, such simulations found their way into essentially all fields of science, business, education, and entertainment. With progress, computerized simulations have become realistic to the point of seeming to be indiscriminate from the reality they depict. Thus, virtual reality can be said to be a natural extension of computerized simulation.

Virtual reality (VR) is a simulated experience that can be similar to or completely different from the real world (Wikipedia). *This definition could subsume all computer simulations, so in this book we restrict the term to mean computer simulations in which the human experiences being a participant in the interaction with the simulated events, and to that extent feels present in that interaction.*

A **virtual environment (VE)** is a computer-generated simulation of a three-dimensional environment that can be interacted with in a seemingly real or physical way by a person using special electronic equipment, such as a helmet with a screen inside or gloves fitted with sensors.

Augmented reality (**AR**) is an interactive experience of a real-world environment where the objects that reside in the real world are enhanced by computer-generated perceptual information, sometimes across multiple sensory modalities, including visual, auditory, haptic, somatosensory and olfactory sensory modalities (Wikipedia). Synonymous with augmented reality are the terms **mixed reality** and **computer-mediated reality**.

AR can be defined as a system that incorporates three basic features: a combination of real and virtual worlds, real-time interaction, and accurate 3D registration of virtual and real objects (Wikipedia).

Taken together, robotic telepresence and virtual/augmented reality are about how people can use certain new technologies to extend their physical and sensory reach to engage with other people or things in new places.

Immersion is the human operator's degree of feeling, compellingly, as though present in the actual remote environment or the virtual environment.

The term **avatar** is used to refer to an icon or figure (a virtual robotic mechanism, human, or animal). It represents a particular person, usually the person controlling the actions of that avatar) in the virtual environment. There can be multiple avatars controlled by multiple persons operating simultaneously in a common virtual environment.

A **metaverse** is a recently coined term (from the 1992 science fiction book *Snow Crash* by Neal Stephenson). It refers to a virtual reality space in which users can interact with a computer-generated environment and other users (Oxford). It typically implies a large-scale network of 3D virtual worlds focused on social connection. In futurism and science fiction, the term is often described as a hypothetical iteration of the Internet as a virtual world that is facilitated by the use of virtual and augmented reality headsets (Wikipedia). There is currently enormous hype being generated by the technology firms Meta (formerly Facebook), Google, and Apple, all of which are competing in this space and making major promises for new technology. Some other large technology firms use the term *metaverse* to refer to the experience that they are marketing in the form of their technology. This author has some skepticism and worry about the new virtual world being hyped, as will become evident in Chapter 4, "Challenges", which is one motivation for writing this book.

Metaverse is not to be confused with *multiverse*, a hypothetical group of multiple universes, as considered by cosmologists to explain the uniqueness of nature on earth. Together, these universes in the multiverse include everything that exists: the entirety of space, time, matter, energy, information, and the physical laws and constants that describe them (Wikipedia).

DIFFERING EMPHASES IN MEANING OF *TELEPRESENCE*

In this book, we interpret *telepresence* broadly, consistent with the first definition above, to include:

(1) *having the sensation* of being present at the actual remote **or** virtual environment;
(2) *giving the appearance* of being present at the actual remote or virtual environment;
(3) *having an effect*, via telerobotics or virtual environment simulation, at the physically remote or virtual location.

Note the different meanings are essentially three different functions. (1) is normally a sensory experience of the primary human operator, a feeling of being "immersed" in the remote (actual or virtual) environment. (2) can be a perception of the primary operator or of other people (he or she is "there" in the remote environment), which does not necessarily include (1). (3) is the perception that a robotic mechanism or avatar is making changes in the remote actual or virtual environment.

The cartoon below may help to see the close relationship between robotic telepresence and virtual reality. It is a cartoon I drew many years ago in writing about robotic technologies that at the time were futuristic but today are directly upon us.

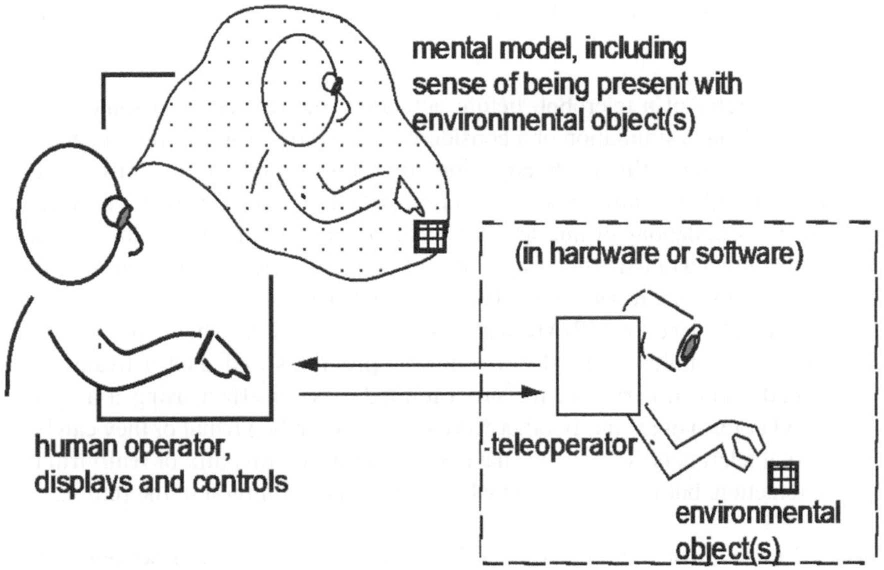

FIGURE I.1 The cartoon illustrates the relations between robotic telepresence and virtual presence. In both cases, a human observes a display and operates controls to manipulate an observed environment, where the environment can be actual but remote or computer-generated. A human operator who is sufficiently mentally "immersed" may not be able to tell whether the environment is virtual or real.

Consider that what is inside the dotted box in Figure I.1 can be video, force, or tactile feedback from an actual robot or other devices in a remote location. Or it can be a *computer simulation* of the remote robot device, where there is *no actual* robot. To the extent that the computer graphics or force/tactile sensing and feedback are of high quality, the human experiences a sense of being immersed in that simulated unreal (virtual) environment. In this case, we have what is called *virtual telepresence*, or more commonly *virtual reality* (VR). The human operator is using hand controls that are suited to the actual or virtual.

In the case of an aircraft or driving simulator, the controls are made to be the same as in the real vehicle, for example. In the case of controlling an actual remote robot, there is an electronic communication channel with characteristics that may pose problems such as time delay or bandwidth constraints. In the case of virtual reality, the electronic communication with the graphic computer image generator is direct. Whether virtual reality or robotic telepresence, the human operator has a mental model of the task, accompanied by some degree of feeling or presence (the term *immersion* is used) in the real remote or the virtual environment. The experience of presence augments the mental model, thus motivating and reinforcing the memory of details and dynamic relationships of what is being observed.

In the future, the relation between robotic telepresence and virtual reality promises to be closer for other reasons:

1) An operator of a telerobot, before activating the telerobot on some task, can activate a simulation of a considered action to test what might happen. To some degree this is already happening, for example, in air traffic control (considering aircraft to be telerobots), where the controllers perform ad hoc simulations of multiple aircraft maneuvering through bad weather and landing while maintaining required separation. Many other varieties of mixing simulation with actual telerobot control are anticipated.
2) A second blend of VR/AR with telerobot control is where the human, operating within a virtual environment, programs and then activates an actual robot to carry out the programmed task, in effect using it as just another form of avatar. In other words, avatars can be virtual or they can be actual telerobots. To date, we have not experienced this mix or real/virtual interaction, but I see no reason why it will not be common in the future.

Finally, I issue a disclaimer: The book is in no way an in-depth description of the technologies of robotic telepresence or virtual reality. And it mostly neglects the relevant technologies of digital communication and artificial intelligence, two disciplines that would require much more space than is available here. But I hope, whatever the level of technology in the reader's background, the book will provide a general understanding and appreciation of telepresence, its potential benefits, and its perils.

1 Early History of Robotic Telepresence and Virtual Reality

The experiencing of a virtual reality is nothing new or inherently bad. Millennia ago, mothers read stories to their children and elders conveyed oral history to families, to motivate their imaginations to experience some form of telepresence, ones that either actually preexisted or never existed. Verbal storytelling was augmented by acting, dancing, and singing, all to stimulate not only the emotions but also the imagination and virtual experience in some form. When listening to a compelling storyteller or reading a good book, one can have the sense of being immersed in a different environment. The very early technological aids were the first implements to draw in the sand or chisel pictographs into stone. Then came pigments to provide color and then crude musical instruments. Gradually, from the Middle Ages until now, technology for more sophisticated forms of drawing, painting, sculpture, and instrumental music emerged. Photography came into being in the nineteenth century, and video and computer graphics in the twentieth century followed.

Computer-generated imagery, whether intended for visual, auditory, or haptic senses, has brought us a new level of control in stimulating the experiencing of virtual reality (VR). Most people would agree that through the ages, motivations for arousing and stimulating the imagination and providing VR experiences have been mostly altruistic or at least benign. Children of any age delight in virtuality, and mostly it is regarded as culturally enriching. Careful and objective evaluation of the use of VR for psychological conditioning, from baseball pitching machines to flight simulators to acting out experiences in psychotherapy, has shown positive benefits. We are motivated, therefore, to continue to enhance our technological capabilities to create VR experiences for our children, our adult trainees, and our patients.

Although robotic telepresence and virtual reality are sometimes considered separate technologies because they are so closely interrelated and their histories overlap, providing separate histories could be confusing.

I can offer some personal recollections of the origins of robotic telepresence and virtual reality, from my own time at MIT and in my own lab. I believe serious technical developments in telepresence began with the advent of master–slave manipulators for remote handling in nuclear "hot labs". The first purely mechanical one was built by Raymond Goertz of Argonne Laboratory in 1948, followed by his electrical model based on servomechanisms with force reflection in 1954. Figure 1.1 shows an example. The possibilities of extending the distance between master and slave, with video feedback added, became obvious. In 1963, the US Navy began the

DOI: 10.1201/9781003297758-2

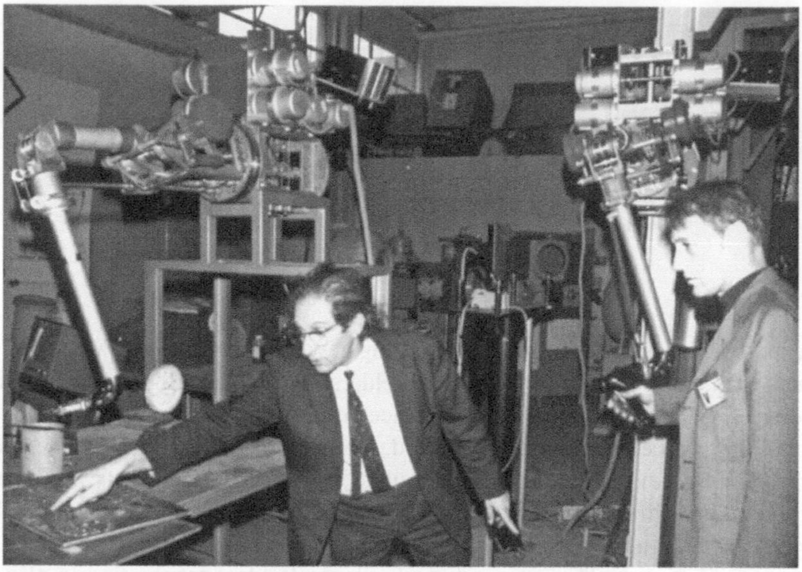

FIGURE 1.1 A six-degree-of-freedom force-reflecting master–slave manipulator system. (Note: "master–slave" is a controversial term in today's world, but no substitute term is in common use.) On the left is Jean Vertut, a French pioneer in robotics, director of a laboratory for the French Atomic Energy Commission, and an author of a 1985 book on Teleoperators and Robotics (Vertut and Coiffet, 1987). The (much younger!) author is on the right.

development of underwater manipulators, leading to undersea exploration, e.g., of the *Titanic*. During those early years, all sorts of terms arose, including *telepuppet*, *telechirics*, *telefactor*, and *cybernetic anthropomorphic mechanism*.

It wasn't until 1980 that Marvin Minsky (Minsky, 1980) popularized the term *telepresence* to emphasize the experience of being present at a remote location (Akin et al., 1983). He saw that teleoperation went beyond just *doing* things at a distance. At about the same time, Susumu Tachi at the University of Tokyo coined the term "telexistence" having the same meaning (Tachi, 2010).

In the early 1960s, NASA funded my Human–Machine Systems Lab at MIT to experimentally explore remote manipulation control with transmission time delay, anticipating lunar rover control. We used a crude manipulator restricted to planar (2 DOF) plus grip. For tasks of different time delays and tolerances, we measured the time required to grasp a block and move it laterally to a new position within a given position tolerance. Results followed a very predictable pattern, as a function of both the time delay and what Fitts (1954) called "index of difficulty". As delay increased, the subject had to make more "moves-and-waits" to achieve the task. Following Shannon's definition of information, Fitts defined the "bits" in a move as log 2 of the move distance divided by the tolerance (Sheridan and Ferrell, 1963)(Figure 1.2).

Ferrell and I suggested a "move-and-wait" strategy for coping with two-way signal transmission delays. That led us to promote the advantages of "human supervisory

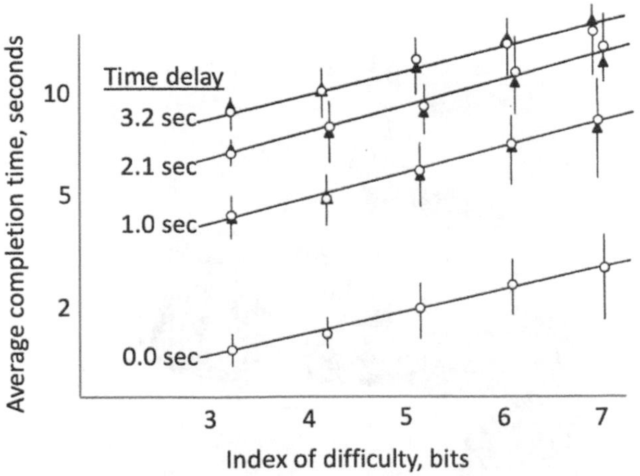

FIGURE 1.2 Sheridan and Ferrell, 1963; Ferrell and Sheridan (1967) results.

control", where a human intermittently programs a computer and the computer performs the feedback control (whether or not there is a time delay). Supervisory control and how it fits into the telepresence picture will be discussed later in the chapter on Elements of the Technology.

Artists got into the VR act early. In 1962, Morton Heilig built a prototype of his immersive, multisensory, mechanical multimodal theater called the *Sensorama* and created five short films to be displayed in it. On August 28, 1962, Heilig was granted US patent 3050870 for a "Sensorama Simulator".

Around the same time, the US Atomic Energy Commission joined NASA in surveying teleoperator and human augmentation developments, resulting in a series of meetings and three reports, NASA SP 5047, 5070, and 5081, spanning 1967–1970. These reports summarized progress in teleoperation up to that time (Johnsen and Corliss, 1967).

In 1968, Ivan Sutherland at the University of Utah developed a stereoscopic head-mounted display pictured in Figure 1.3. He also spoke of having a room with digital displays on all six sides of the 3D space, corresponding to the video images recorded from elsewhere. This display configuration is now called a "cave". Sutherland is often considered to be the founder of virtual reality hardware and software.

In 1977, David Em, an artist at NASA's Jet Propulsion Lab, set up a "navigable virtual worlds" lab (Em and Tractman, 1988), to be followed by a 1978 MIT project called the Aspen Movie Map, which enabled users to wander the streets of Aspen, Colorado (Lippman, 1980). In 1979, Eric Howlett designed what he called the large expanse extra perspective (LEEP) optical system, which was later redesigned to become NASA Ames Research Center's first VR workstation under Scott Fisher (Fisher, 1986).

In 1982, Tachi and Abe published "Study on Telexistence", using their newly coined term for telepresence. In 1983, the Japanese established their "Advanced

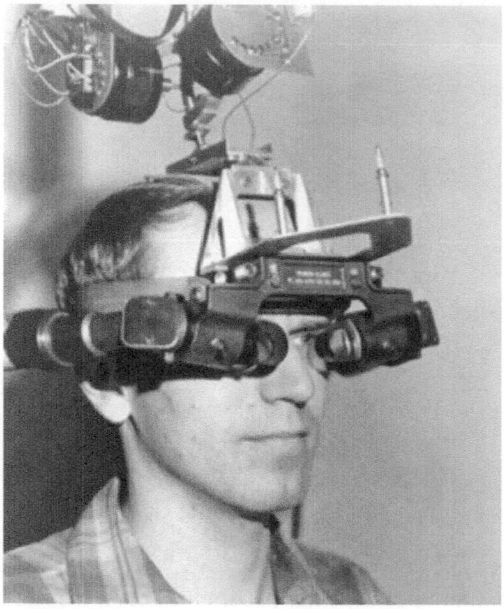

FIGURE 1.3 Ivan Sutherland's 1968 stereoscopic head-mounted display. It was so heavy it had to be suspended from the ceiling (NASA).

Robot Technology in Hazardous Environments" project, which funded a set of major Japanese projects in telerobotics, largely from Tachi's laboratory at the University of Tokyo. In 2010, Tachi authored a book titled *Telexistence* which nicely summarizes the extensive telerobotic technology developed in his lab to that point.

The widespread use of the term *virtual reality* (VR) can be traced to computer software engineer Jaron Lanier. He designed some business-grade VR at his firm VPL Research. In the late 1980s, Lanier's company developed an early *dataglove* to provide a crude tactile display of digital information. This author patented a touch display in 1986 that converted tactile (force pattern) to a light pattern (feeling by seeing). A year later my student Sam Landsberger patented a robot configuration based on the Stewart platform kinematics that is used to provide roll, pitch, and yaw in flight simulators (see Landsberger, 1987, in the chapter on Elements of the Technology).

In the late 1980s, NASA funded my MIT colleague Nathaniel Durlach and me to bring NASA teleoperator engineers together with virtual reality software folks like Jaron Lanier at a waterfront hotel in Santa Barbara. The aim was to explore the possibilities of integrating VR developments with those in telepresence. It was clear, as suggested in the cartoon of Figure I.1, that from the human operator's perspective, VR and telepresence control of actual objects can look the same to the human operator and present the same issues, whether the "remote" action is generated by software or is mechanically real. Resulting from that meeting, the MIT Press founded the journal *Presence: Teleoperators and Virtual Environments*. Its first issue was published in 1992 with Durlach and me as editors, and it continues bimonthly to this day. This journal has been the primary publication for studies on the psychological aspects of virtual reality and telepresence and will figure heavily in the Bibliography of this book. At the same time, my book *Telerobotics, Automation and Human Supervisory Control* (Sheridan, 1992) reviewed laboratory research to date in my own lab and elsewhere, and attempted to sort out the problems.

In 1985, the *Titanic* was discovered by a crew from the Woods Hole Oceanographic Institute. Figure 1.4 shows the tethered teleoperator Jason Junior peering into a portal on the *Titanic*, while the operators on the surface observed what it saw with its video. It later swam around inside. This was a clear early example of robotic telepresence.

In 1990, Eric Gullichsen founded Sense8 Corporation. It developed what was called the World Tool Kit, which allowed real-time graphics with texture mapping on a PC (see Stampe et al., 1993).

In 1992, inventor Louis Rosenberg, at the US Air Force Armstrong Lab, built a system that included a stereoscopic image display from the remote environment and a crude touch feedback arrangement as part of an upper body exoskeleton. By that time, NASA (Steven Ellis and his colleagues at NASA Ames) were performing virtual reality experiments with head-mounted displays (Ellis et al., 1985).

Aircraft pilot training simulators are a specialized form of virtual reality and have become highly developed with computer-generated displays of the runway for takeoff and landing and of mountains and other terrains for cross-country flying. Realistic driving simulators have also emerged. They were not originally called virtual reality, but they truly are. They simply preceded the popular terminology. The aim has been to make the simulator flying or driving experience match the real world as much as possible, giving the trainee a real sense of presence as if immersed in the real-world task, but without the risk. These simulators will be discussed further in the Applications chapter.

FIGURE 1.4 The tethered Jason Junior peering into a portal on the sunken *Titanic* (US Navy).

A teleconferencing research project began in 1990 at the University of Toronto, with international partners in four countries. Teleconferencing is probably the form of telepresence technology that has expanded most rapidly. Meetings of large numbers of people, which include real-time drawing and physical demonstrations (as in lecturing), can now be conducted via Zoom and similar technology. Guided tours of museums or historical sites anywhere in the world can be accomplished by using VR platforms such as Facebook's Oculus. We review that technology in Chapter 4.

By 1990, it had become abundantly clear that there is little technologically that impedes human control of mechanical things at any distance, given sufficiently high-quality visual, auditory, tactile, or haptic feedback communication, and given sufficiently fast, accurate, and powerful vehicle movement or mechanical handling for the remote side. By 1990, one could also engage in crude but effective teleconferences over the fast-developing internet. It was also clear that computer-generated virtual reality had immense promise in entertainment and possibly education, waiting primarily on advances in computer and graphical display speed and cost.

The idea of robotics used for surgery began more than 50 years ago, but actual use began in the late 1980s with Robodoc (Integrated Surgical Systems, Sacramento, CA), the orthopedic image-guided system developed by Hap Paul and William Bargar (Paul et al., 1992) for use in prosthetic hip replacement. Intuitive Surgical was founded in 1995 by Frederick Moll, Rob Younge, and John Freund, with help from Kenneth Salisbury, a robot designer who had moved from MIT to Stanford. The company licensed telepresence surgical technology from SRI (Menlo Park, CA) and began developing what would ultimately become the DaVinci Surgical System, the first of which was installed in late 1998.

The DaVinci surgical telerobot, finally approved by the Food and Drug Administration in 2000, is now in wide use, and the distances between surgeon and patient are being extended as needed. It is discussed further in Chapter 4. The need to transfer some mass from the operator's location to the remote site is, of course, an exception: electronic communication does not transfer mass. In the field of medicine, the traditional in-person diagnostic functions of inspection (seeing), auscultation (listening), and palpation of the patient can now be accommodated by telepresence technology, and we are certainly witnessing an increase in other types of telemedicine.

Augmented reality (AR) is really a complement to virtual reality, requiring a modification to the head-mounted display so as to combine virtual images with images from the actually present environment. This is discussed further in Chapter 4. The earliest functional AR systems were in 1992 from the Air Force Armstrong Laboratory. AR promised a variety of applications such as overlaying text or graphics on top of actual current images so that a worker can identify, get instructions about, or learn a current task being observed. This is being applied to a range of activities such as machine maintenance, brain surgery, and museum visitation.

In recent years, the major high-tech firms such as Facebook (Meta), Google (Alphabet), Apple, Microsoft, and Amazon have invested heavily in VR and telepresence technology, and we have yet to see what new forms the technology will take.

REFERENCES

Akin D, Minsky M, Thiel E, et al.: *Space Applications of Automation, Robotics and Machine Intelligence Systems (ARAMIS)*, Phase 2. Volume 2: Telepresence project applications. NASA CR 3735, 1983.

Ellis S, Kim, W, Tyler M, et al.: Visual enhancements for perspective displays: Perspective parameters, in *Proceedings of the 1985 International Conference on Systems, Man and Cybernetics*, New York, IEEE, pp. 815–818, 1985.

Em, D and Trachtman P: An impressionist with a computer, *Smithsonian Magazine*, 1988.

Ferrell W and Sheridan T: Supervisory control, *IEEE Spectrum*, October 1, 1967.

Fisher S: Virtual interface environment, in *IEEE/AIAA 7th Digital Avionics Systems Conference*, Fort Worth, TX, October 13–16, 1986.

Fitts P: The information capacity of the human motor system in controlling the amplitude of movement, *Journal of Experimental Psychology: General* 47(6): 381–391, 1954.

Johnsen E and Corliss W: *Teleoperators and Human Augmentation*, NASA SP-5047 1967; *Teleoperator Controls, NASA SP-5070 1968; Advancements in Teleoperator Systems*, NASA SP-5081, 1970.

Lippman A: Movie-maps: An application of the optical videodisc to computer graphics, in *Proceedings of the 7th Annual Conference on Computer Graphics and Interactive Techniques*, Seattle, Washington, DC, pp. 32–42, 1980.

Minsky, M: Telepresence. *Omni*, June, 45–51, 1980.

Paul H, Bargar W, Mittlestadt B, et al.: Development of a surgical robot for cementless total hip arthroplasty, *Clinical Orthopaedics and Related Research* 285: 57–66, 1992.

Rosenberg L: The use of virtual fixtures as perceptual overlays to enhance operator performance in remote environments, technical report AL-TR-0089, Wright-Patterson AFB, OH, USAF Armstrong Laboratory, 1992.

Sheridan, T: *Telerobotics, Automation and Human Supervisory Control*, Cambridge, MA, MIT Press, 1992.

Sheridan T and Ferrell W: Remote manipulation control with transmission delay, *Transactions IEEE* HFE-4: 25–28, 1963.

Stampe D, Roehl B and Eagan J: *Virtual Reality Creations: Explore, Manipulate, and Create Virtual Worlds on Your PC*, Victoria, Australia, Waite Group Press, p. 97, 1993.

Sutherland, I: A head-mounter three dimensional display, *AFIPS '68 (Fall, part I): Proceedings of the* Fall Joint Computer *Conference*, part I, December, pp. 757–764, 1968.

Tachi S: *Telexistence*, Tokyo, World Scientific, 2010.

Tachi S and Abe M: Study on telexistence, design of visual display, in *Proceedings of the 21st SICE Annual Conference*, pp. 167–168, Tokyo, July 28–30, 1982.

Vertut J and Coiffet P, *Teleoperation and Robotics: Applications and Technology*, New York, Springer, 1987.

2 Elements of the Technology

This chapter takes a quick look at the major technologies that support robotic telepresence and virtual reality (VR)/augmented reality (AR). The sections include Robot Configurations; Visual Displays and Video Cameras; Haptics and Smart Gloves; 3D Audio; Control for Telepresence Systems; and Telepresence Communications. All apply in one way or another to *both* robotic telepresence and virtual/augmented reality.

TELEROBOT CONFIGURATIONS

What do robots look like? Robots come in a myriad of configurations: the more the degrees of freedom or DOF (number of possible independent motions), the greater the number of different robot configurations. This field is called *kinematics*. Apart from the *kinematics* per se, there are many different appearances. These are determined by whether the robot's designed function is for factory assembly lines or warehouses, as tools for precise surgery, as accessories for firefighting or police surveillance, as toys, or as weapons of war. Catalogs of commercial robots are available on the internet, for example, as sponsored by the Institute for Electrical and Electronics Engineers (IEEE).

Alternative kinematics. Figure 2.1 shows alternative kinematics for three degrees of freedom (DOF), different combinations, or translation and rotation. If the lower right cylinder is the base, one can easily determine that three more DOF are needed to place the other endpoint at any location at any angle. To avoid collisions with objects in the robot workspace, additional degrees of freedom are needed to "reach around". The master–slave manipulator shown in Figure 2.2 has six DOF. It should be noted that the above comments relate to rigid bodies. New developments include extended tube robots, snakes, and robots made of soft materials.

There is a strong connection of telerobot kinematics to telepresence in terms of the human operator being able to make an identification between the robot arm and his own arm, especially with regard to roll, pitch and yaw of the gripper or "end effector". When the operator can only control the rate of each independent DOF (and not the endpoint position) that identification is easily lost. One can assert that naturalism (and hence speed and error avoidance) corresponds to the degree to which the operator's mental model matches the kinematics of the robot. That kinematic correspondence has a strong influence on telepresence.

A six-degree-of-freedom robot with all rotational DOF is shown in Figure 2.2. The rotational joints are identifiable by what appear to be metallic bands.

DOI: 10.1201/9781003297758-3

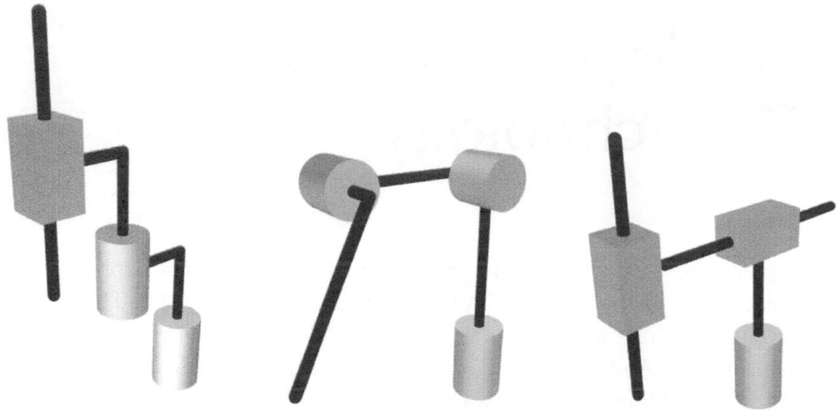

FIGURE 2.1 Alternative kinematic arrangements for three DOF robot arms (Wikimedia Commons). The cylinders indicate rotation, and the rectangular blocks indicate translation.

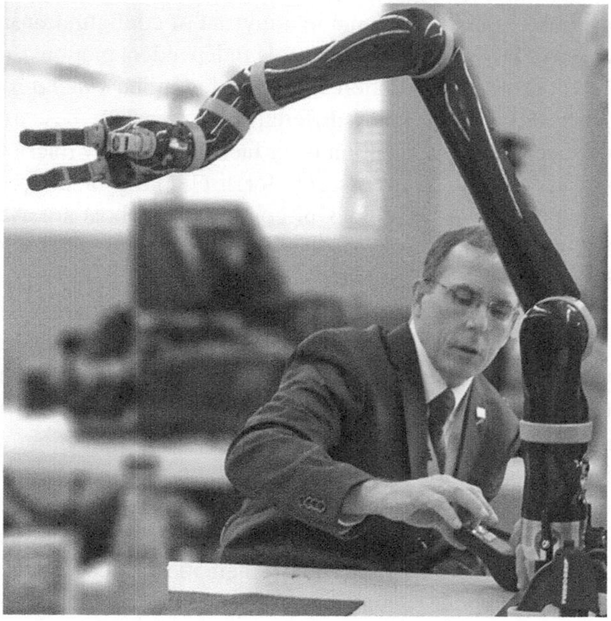

FIGURE 2.2 A six DOF rotational robot arm (Wikimedia).

A robot configured on the basis of Stewart platform kinematics is shown in Figure 2.3. Six cables are maintained in tension by a single compression spring. By independently adjusting the six cables, the upper platform can be adjusted, within a limited range, in any of the six DOF. A patent by Landsberger and Sheridan (1987) demonstrated a combination of speed and force that easily exceeded that of many conventional robots, partly because of the lack of mass of the arm elements of conventional robots.

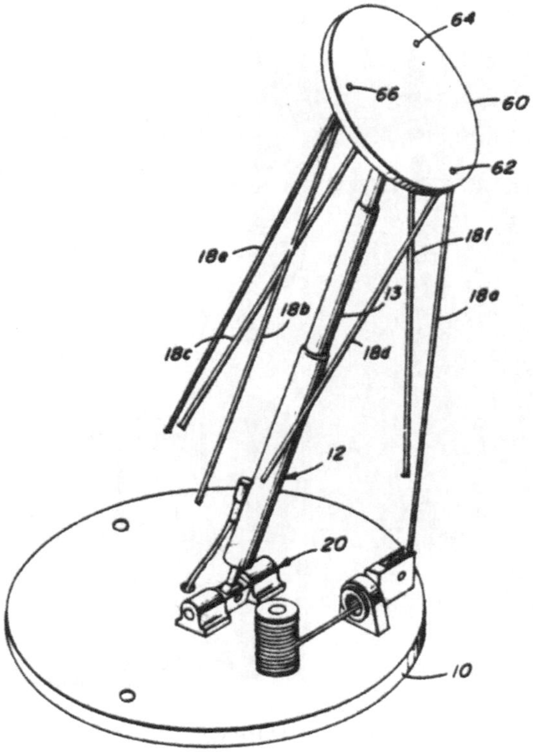

FIGURE 2.3 The Landsberger 1987 robot. (Landsberger et al, US Patent 4666362, Parallel Link Manipulators.)

Anthropomorphic configurations. Anthropomorphism is the degree to which a telerobot's kinematic configuration resembles a person, with a "head", two arms on a torso, and a vertical body with some means for mobility on the bottom. Anthropomorphism has the advantage that a human operator can identify with the robot. That is particularly true if the operator is functioning in a master–slave control mode. An anthropomorphic robot is pictured in Figure 2.4. This one happens to be the winner of the DARPA (Defense Advanced Research Projects Administration) contest which involved performing a number of real-world tasks including picking things up, opening a door and passing through an entry, and adjusting a valve (as shown in the figure). Some say that dressing the telerobot in human clothing will enhance the device by making it an easily controllable avatar for the human operator.

A telepresence robot for VR meetings. In stark contrast to articulated six or seven DOF robot arms designed to let users pick up things and manipulate the environment, one popular form of telepresence robot shown in Figure 2.5 has no arms. Its only function is to move a camera, display, microphone, and speaker in two DOF to any surface location and to rotate to any angle. This simple and inexpensive configuration permits a user to attend a meeting remotely, listen in, and converse with someone nearby, with each person seeing and hearing the other. There is no need

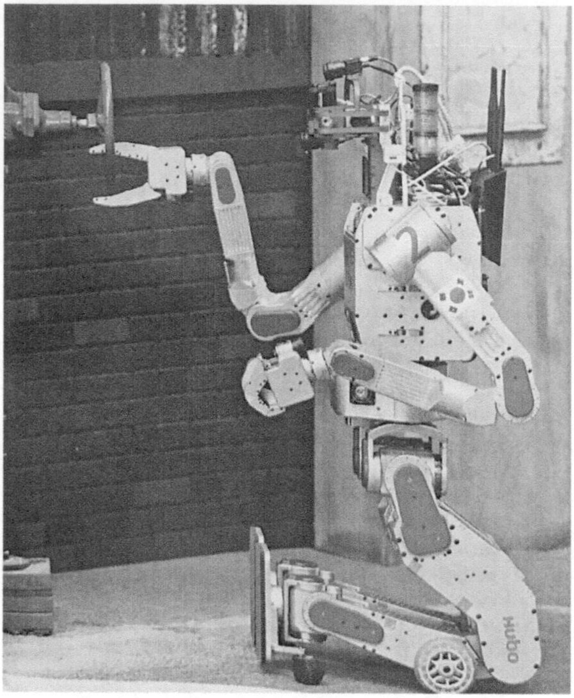

FIGURE 2.4 An anthropomorphic robot (DARPA).

for manipulation. Such devices have been used, for example, in shops, airports, and other public spaces, to greet people and respond to questions. A live person can be at the other end, or with sufficiently sophisticated speech recognition and artificial intelligence (AI) computation, the device can do its job autonomously.

It can be said that any machine that includes a sensor, a computer, and a motor and is remotely operated is a telerobot. A modern commercial aircraft and a modern automobile fit this pattern, though they are not anthropomorphic in form, as do a myriad of automatic devices operated over the "internet of things" such as home heaters, stoves, and dishwashers.

VISUAL DISPLAYS AND 3D VIDEO CAMERAS

Head-mounted displays. For robotic telepresence, to achieve a strong sense of presence at the remote site, the robot needs a video camera mounted on the robot's "head", that tracks the human operator's head in all degrees of freedom including lateral translation (though three rotations of pan, tilt, and yaw usually suffice). Then the human sees what the robot sees when looking in the same direction through a head-mounted display (HMD). The camera should be of high resolution and color.

In either robotic telepresence or VR viewing, the rotation and movement sensors in the head-mounted display record the human operator's view direction and lateral

FIGURE 2.5 A telepresence robot with surface mobility, a display, and a microphone for meetings and greetings (Wikimedia Commons).

movement. This guides the remote robot "head" camera (and selects the VR view) corresponding to the HMD direction and location. This is a key feature of what provides the sense of telepresence in both the actual remote and the virtual environment.

There have been many HMD designs, both developmental (laboratory prototypes) and commercial. But today, there are three that have particularly captured the market. All three can serve in viewing either robotic telepresence or the virtual reality.

Shown in Figure 2.6 is the Meta (Oculus 2) device. It consists of two conventional computer displays, including a full self-contained computer, offset with suitable optics for the two eyes. It also comes with left- and right-hand controls that provide for pointing and selecting objects within the view. Advertised specifications are given below.

Oculus Quest 2 (VR HMD) Specifications

Cost \$300–\$400
1832 × 1920 pixels per eye
Screen refresh rate 90 Hz
Weight 503 g
6 DOF tracking

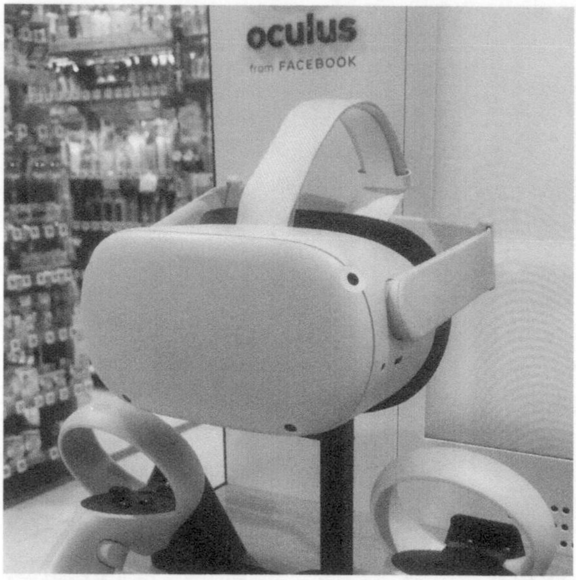

FIGURE 2.6 Meta's Oculus head-mounted display with hand controls below (Wikimedia Commons).

CPU: Qualcomm® Snapdragon XR2 Platform
Memory 6 GB
Integrated audio
6 DOF independent hand-held pointers
9 × 9 ft play space

Figure 2.7 shows the front of the HTC Vive Cosmos, one of several models made by the Taiwanese firm HTC. It is similar to the Oculus in many ways.

HTC Vive Cosmos Specifications

Cost $800–$1000
Pixels per eye 2880 × 1700
90 Hz LED
110-degree field of view
Controllers tracked by six cameras
Weight 702 g

A third HMD is Microsoft HoloLens shown in Figure 2.8. It does not fill the screen with a computer-generated image as do the Oculus and HTC displays. Instead, it provides a limited-size computer-generated holographic image, generated through "waveguides", making it appear 3D. It thus enables the operator to reach out and

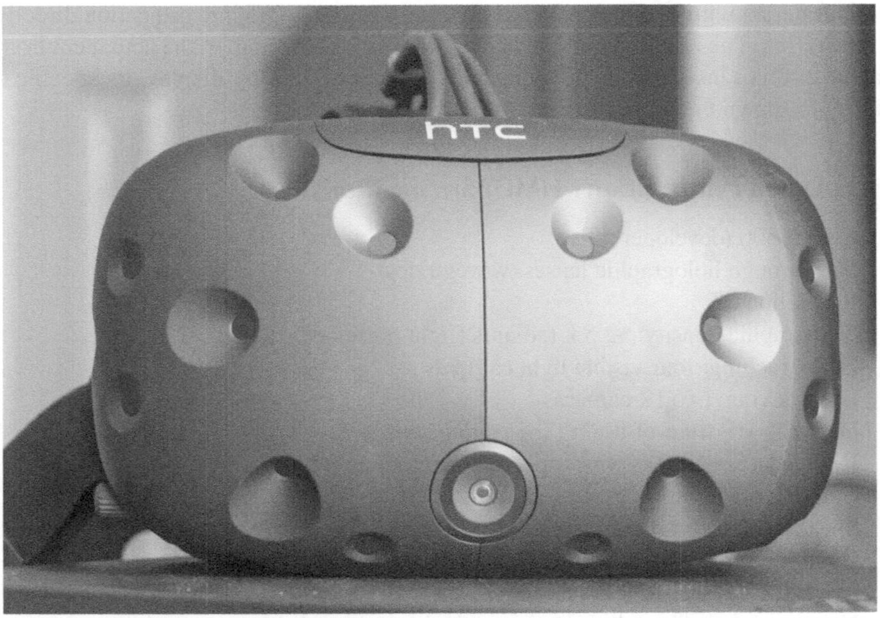

FIGURE 2.7 HTC Vive Cosmos headset (Wikimedia Commons). Note the cameras pointing in many directions.

FIGURE 2.8 Microsoft's Hololens HMD for augmented (mixed) reality. Wearer can see and "touch" a virtual object (Wikimedia Commons).

touch/interact with that image, based on laser-based holographic projection directly to the retina. The remainder of the field of view is see-through to the actual environment. In this sense, it is an "augmented" or "mixed" reality display. Its specifications are shown below.

HoloLens 2 (Mixed Reality HMD) Specifications

Cost $3500 (developer only)
See-through holographic lenses (waveguides)
Resolution 2K
Holographic density: >2.5 k radiants (light points per radian)
Head tracking: four visible light cameras
Eye tracking: two IR cameras
Depth: 1-MP time-of-flight (ToF) depth sensor
IMU: accelerometer, gyroscope, magnetometer
Camera: 8-MP stills, 1080p30 video
Speakers: built-in spatial sound
Hand tracking: two-handed fully articulated model, direct manipulation
Eye tracking: real-time tracking
Voice: command and control on device; natural language with internet connectivity
Windows Hello: enterprise-grade security with iris recognition
Storage: 64 GB UFS 2.1
Windows Holographic operating system

Cave VR display (sometimes regarded as an acronym for "Cave automatic virtual environment"). The cave configuration is simply a set of six screens in a room (originally projected, but they now can be LED displays) as shown in Figure 2.9. The human viewer stands or sits in the center of the room, and can turn his body and look in any direction. The appropriate VR image is already there, consistent with the human's head position and orientation, provided that the robot in the latter case includes a camera looking in all directions (and thus providing an image on all sides of the cave).

Cameras for generating VR environments. The omnidirectional VR records in many directions as the camera is moved through some environment. This naturally poses a huge computer memory requirement. For example, National Geographic (2020) offers telepresence tours of Notre Dame Cathedral and Machu Picchu that take an HMD wearer on a VR tour. The equivalent head position for walking through the virtual environment is fixed by how the omnidirectional camera was originally moved through the actual environment.

3D virtual reality images are normally generated by simultaneously taking continuous video images from multiple directions as the camera moves through the planned VR environment. If, however, viewing the virtual environment in all possible directions and *from many possible positions* is to be available, then many separate cameras will be necessary, viewing from different positions (Figure 2.10). The

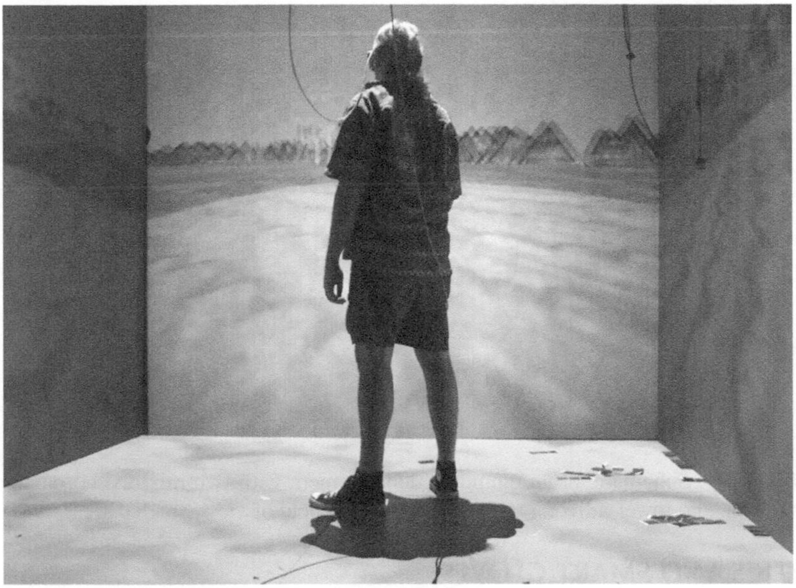

FIGURE 2.9 A cave display: Every wall, floor, and ceiling has an image that corresponds to the scene as viewed from the person's perspective (Wikimedia Commons).

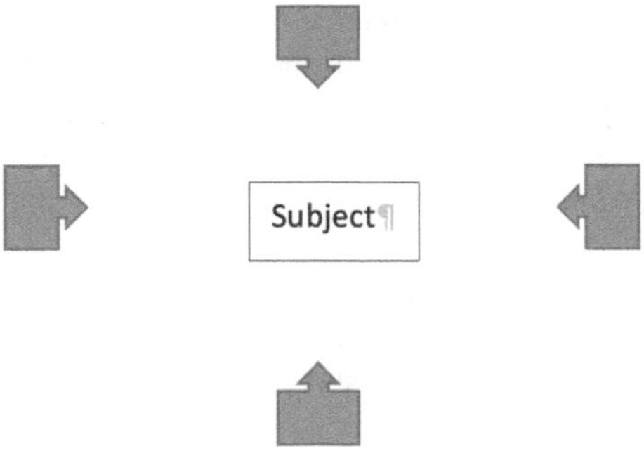

FIGURE 2.10 Picturing an object simultaneously from multiple directions.

separate images are later electronically merged (as is now common in conventional 360-degree lateral panoramic viewing).

For viewing in the virtual environment from one position (and if looking straight up or straight down is not to be available), a single flat video sensor with mirrors, such as is diagrammed in Figure 2.11 is adequate, providing a panoramic field of view, such as is shaded in the diagram. In this case, the distorted image must then be unscrambled electronically to be consistent with 360 degrees or omnidirectional rotation of the HMD. One such optical arrangement is shown in Figure 2.11.

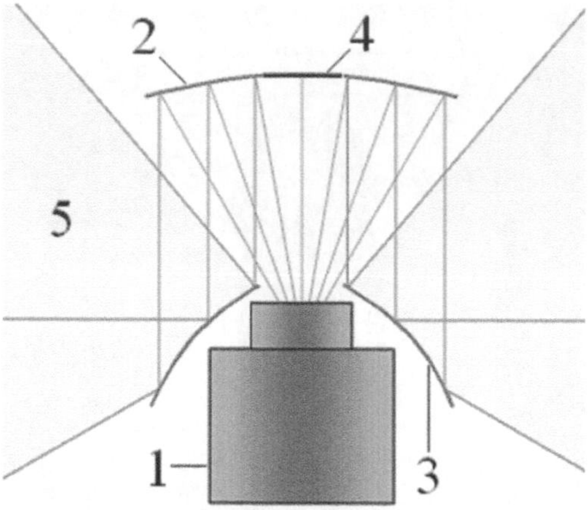

FIGURE 2.11 Schematic of an omnidirectional camera with two mirrors: (1) camera, (2) upper mirror, (3a) lower mirror, (4) black spot, and (5) field of view (shaded area).

HAPTICS AND SMART GLOVES

Robot haptics. By its etymology, haptics refers to touch but has come to be associated with senses of body motion (kinesthesis), body position, and gross forces on the muscles and tendons. Haptic sensing can be either by a robot or by a human. Both are necessary for haptic telepresence,

Touch patterns are important for interactive mechanical applications, especially when the interactions are unpredictable. Robots use touch signals to map the profile of a surface, especially critical in hostile environments. Four measurable properties—force, contact time, repetition, and contact area change—are ways to categorize touch patterns. A major challenge is to map the individual components of the touch pattern.

Since some touch signals result from the robot's own movements, it is important to identify the external tactile signals for accurate operations. Electronic logic that relies on the prior knowledge of signal statistics can enable the robot to predict the sensor signals caused by the robot's own internal motions, thus screening out false signals.

Robotic touch sensing can be based on many principles: capacitors, piezoelectric material, strain gauges, color changes, optical intensity, etc. There are two types of capacitive sensing systems: (1) mutual capacitance, where the object (finger, conductive stylus) alters the mutual coupling between row and column electrodes, which are scanned sequentially, and (2) self- or absolute capacitance, where the object (such as a finger) loads the sensor or increases the parasitic capacitance to ground (Figure 2.12). Currently, there are electrically conductive polymers that are useful in touch sensors.

An early touch sensor/display patent from the author's lab is shown in Figure 2.13 (Sheridan, 1986). A flexible material has holes that are co-located with optical fibers, all spaced very closely together. A touch (force) pattern is then displayed to the human operator by the light pattern emanating from the other end of the fiber bundle.

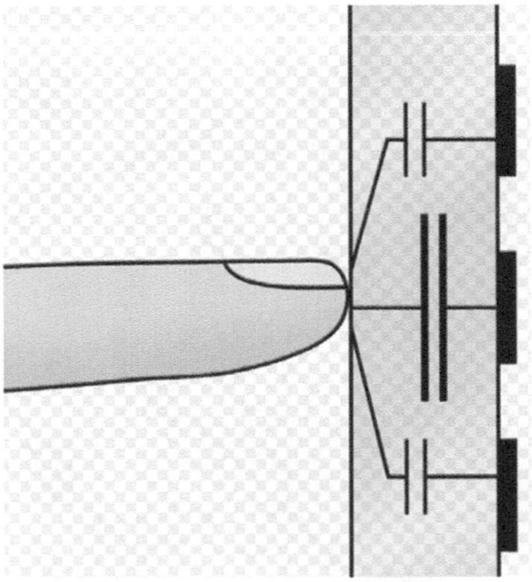

FIGURE 2.12 Touch measurement by electrical capacity.

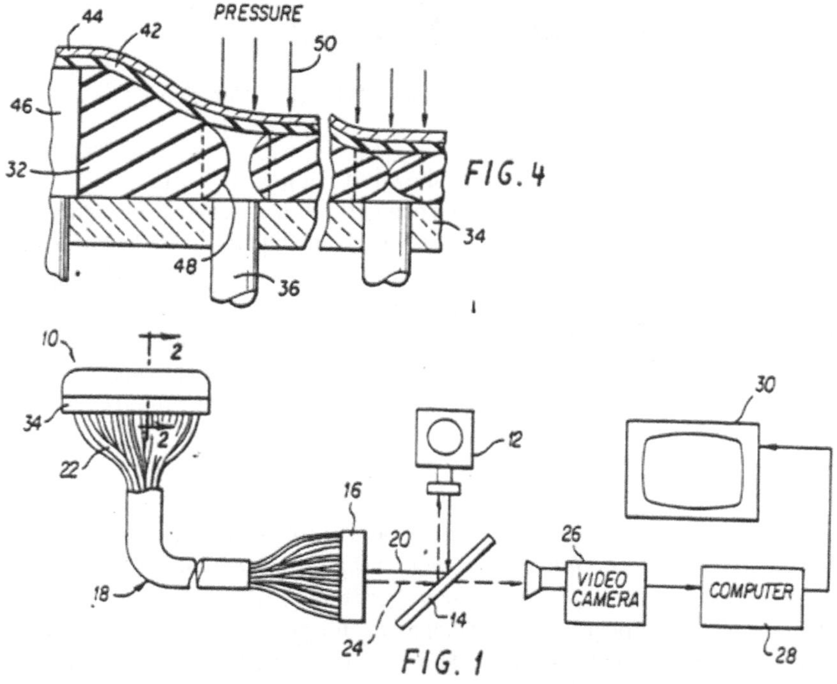

FIGURE 2.13 Patent drawings for an early robotic touch sensor/display. External pressure closes the spaces so that less light is reflected off a deformable mirror into an optical fiber. (Sheridan et al, US Patent 4599908, Opto-mechanical touch sensor.)

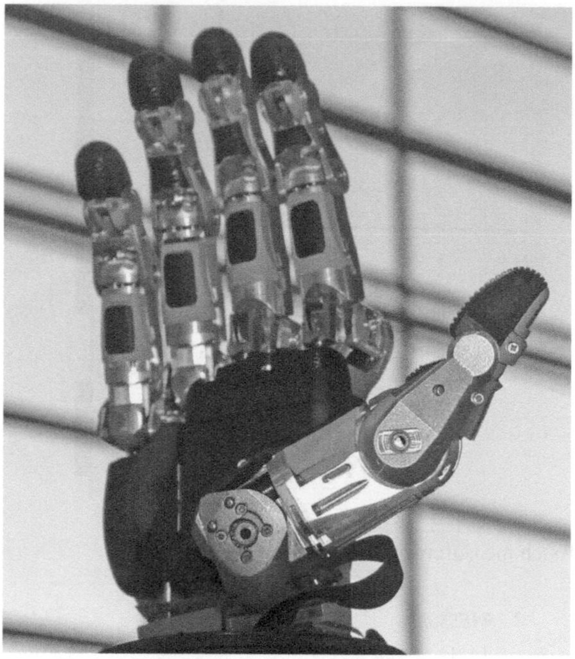

FIGURE 2.14 A robotic hand with touch sensors (Wikimedia Commons).

A robotic hand with touch sensors is shown in Figure 2.14.

When controlling a mechanical arm/hand, the touch sensors are activated by grasping actions of the hand, presumably relaying touch patterns to the human operator. But the robot arm can also provide overall force feedback, which can be fed back to the human operator's muscles and tendons.

Master–slave manipulator force feedback is achieved by measuring the current in the slave motor for each degree of freedom and using that current to drive a motor in the opposite direction. The master–slave manipulator system pictured earlier operated that way.

Smart gloves are gloves worn by the human operator in either case: (1) telerobotic control, *to measure haptic pose and/or receive haptic feedback information* corresponding to the palm and each finger from the robotic hand, or (2) virtual reality interactions *to measure haptic pose and/or simulate touch and force patterns on the fingers and hand as a function of virtual grasping and pushing, etc.*

Smart gloves have been under development during the last 40 years to support human–computer interaction based on hand and finger movement. Despite the many devoted efforts and the multiple advances in related areas, these devices have not become mainstream. Nevertheless, during recent years, new devices with improved features have appeared, being used for research purposes too. Caeiro-Rodriguez et al. (2021) provide an extensive review of current commercial smart gloves focusing on three main capabilities: (1) hand and finger pose estimation and motion tracking,

(2) kinesthetic feedback (whole hand), and (3) tactile feedback. For the first capability, a detailed reference model of the hand and finger basic movements is proposed. The paper reviews the technologies involved and the main applications, and it discusses the current state of development. Reference models to support end-users and researchers comparing and selecting the most appropriate devices are identified as a key need. Many companies have different versions of their gloves: Cyberglove/ Cybergrasp, 5DT, Dexmo, Synertial, Manus, Nansense, Noitom, SenseGlove, and VMG. In many cases, they provide the same glove with a different number of sensors to enable the tracking of more degrees of freedom.

Whole body suits. Sometimes, especially in VR situations, but also in robotic aids for physically disabled persons, it is desirable to measure the position and motion of the whole body, or at least of the upper torso. One way to make these measurements is to integrate the signals from inertial sensors placed strategically on the limbs. Another is for the human to wear a body suit with stretch sensors at the joints or with light reflectors along the limbs such that a video camera can pick up the signals and infer the total body configuration.

3D AUDIO

The human brain estimates the 3D location of a sound source by comparing sounds received at both ears (*difference cues* or *binaural cues*). Among the cues are time differences of arrival and intensity differences. The brain also makes use of monaural cues that come from the interaction between the sound source and the human anatomy, in which the original source sound is modified before it enters the ear canal for processing by the auditory system. These modifications are used by the brain to discriminate the location of the sound source.

Audiologists measure the response of each ear to sound impulses generated at each location around the head. This impulse response is termed the *head-related impulse response* (HRIR). They use these impulse responses to compute the sounds at the ear locations through convolution (overlaying impulse responses for each discrete value of the sound source). This is a standard technique for determining the response to any linear system.

But what about sources directly above the head, such that sounds in each ear are the same? Or what about the situation where the HMD wearer is hearing through earphones, and so cannot benefit from the HRIR effect to locate sounds? Armed with the impulse responses, the computer can generate sounds in the earphones that duplicate what would occur at any location without the earphones. Thus, in a VR environment where the human is wearing earphones as well as an HMD, the computer can create the effect of an insect buzzing overhead (Wenzel, 1993)

Laser-based 3D scanning. Three-dimensional laser scanning is used in many applications to determine the distances from the location of the scanning instrument to any and all other objects in its surround, producing what is called a *point cloud*. Laser scanning is a key sensor technology on the rooftop of many self-driving cars. It is widely used in the architecture and construction industries to measure distances, where electronic timing is so accurate that it can measure the time from a pulse of light to its return. By scanning an object from many different locations of the scanner and at many different angles relative to the object being scanned, and then having software weave together the distances from the scanner to the object, a computer can build up a model of the surface of the object.

CONTROL FOR TELEPRESENCE SYSTEMS

Continuous manual control. Key to telepresence, especially in terms of the third definition, *having an effect*, via telerobotics, at a remote location, is *control*. Direct human control is usually called manual control and is understood in terms of feedback control theory. I think we can explain here the most elementary principle of feedback control, namely having a high gain (sensitivity to error) but at the same time avoiding loop instability.

Consider the block diagram of Figure 2.15, depicting any closed-loop control system, such as a pilot (Controller) flying an aircraft (Vehicle). The reference variable **r** might be a lateral position requirement, such as landing on a runway, and the output might be the actual (new) lateral position **n** relative to the runway.

The block labeled *Vehicle* represents the vehicle dynamics (differential equations) and so too with the other blocks. Each block represents a functional relationship between equations (called a *transfer function*), for the input to or output from that block.

For readers comfortable with a bit of math, it may be useful to derive the most basic principle of why feedback control works: Insofar as the transfer functions are linear dynamic relationships (and couched as Laplace transforms, not to be explained here), their effects are multiplicative. Thus, one can follow the algebra and say that

$\mathbf{n} = \mathbf{V} \times \mathbf{C} \times \{\mathbf{r} - \mathbf{F} \times \mathbf{n}\}$ where × indicates multiplication, so that

$$\mathbf{n} = \{\mathbf{V} \times \mathbf{C} \times \mathbf{r}\} - \{\mathbf{V} \times \mathbf{C} \times \mathbf{F} \times \mathbf{n}$$

$$\mathbf{n} + \mathbf{V} \times \mathbf{C} \times \mathbf{F} \times \mathbf{n} = \mathbf{V} \times \mathbf{C} \times \mathbf{r}$$

$$\mathbf{n}(1 + \mathbf{V} \times \mathbf{C} \times \mathbf{F}) = \mathbf{V} \times \mathbf{C} \times \mathbf{r}$$

so that the ratio of output to input equals

$$(\text{output } \mathbf{n}/\text{reference } \mathbf{r}) = (\mathbf{V} \times \mathbf{C})/(1 + \mathbf{V} \times \mathbf{C} \times \mathbf{F})$$

Insofar as the feedback **F** transfer function is perfect, i.e., F is equal to 1, and the controller **C** is very sensitive, i.e., is a large gain (much greater than one), **V × C** is very large relative to 1, and so (output **n**/input **r**) approaches 1, i.e., perfect control.

This is an ideal case. In reality, there are many factors that may produce instability, including the complexity (order, nonlinearities) of the loop dynamics and time delays within the loop, such as those for communication. One requirement is to keep the loop gain less than unity at any frequency causing instability (where the feedback at that frequency will be positive and thus cause instability). Further consideration of what is required for an acceptable transient response as well as frequency response is beyond what we consider here. The latter issues are discussed in any reference book on control theory.

The diagram and the relations above apply just as easily to steering an automobile, where the reference variable **r** is the lateral (center) position on the highway lane, and the output **n** is the actual position in response to each wiggle of the steering wheel. It is intuitive that the driver's sensitivity to error (controller gain) will make for tightened adherence to staying centered on the lane. These ideas apply to any human control, including the tele-control of a robot.

What can be said of actual human control based on years of research? A cadre of researchers supported by the US Air Force during the 1950s (disclosure: the author was one of them) sought to provide a differential equation model for the human operator in tracking tasks. Tracking in this case means the human continuously moves a hand control to cause a corresponding position or rate change to a vehicle or other device. The application of interest to the Air Force was that the pilot (*Controller*) in those days was "in the loop" with the dynamics (*Vehicle*) of the aircraft, particularly fighter aircraft, and it was critical to understand the human equation to avoid loop instability. (Commercial aircraft these days are seldom flown manually; they are programmed to go from one waypoint to another, climb, or descend to a different altitude, etc.)

Research showed that the human controller is very flexible and is able to adapt to the dynamics of the aircraft. From human-in-the-loop simulation experiments, it became clear that it was much simpler to model the combined *Controller* and *Vehicle*, assuming that the *Feedback* transfer function was essentially 1. The result was surprisingly simple. Across a range of aircraft dynamics, the combined human–aircraft transfer function turned out to be a single integration with the time delay (due to human neuromuscular constraints) of about 0.25 sec (Sheridan and Ferrell, 1974). This has come close to applying to response times for other manual tasks.

In robotic telepresence, sometimes (e.g., humans on Earth controlling telerobots in space) there are pure time delays greater than a small fraction of a second between when a movement is commanded and when there is visual or force feedback. Then, at some frequency (e.g., caused by noise), the feedback, instead of making the error smaller, will make it larger. In this situation, if the gain is large enough, the error at that frequency will grow continuously, causing instability. This problem is much more sensitive to force feedback than to visual feedback, because with visual feedback the brain can more or less ignore the high-frequency jitter, whereas the hand cannot ignore force feedback. If the controlled process has large inertial lags, then that dampens the loop gain at the higher frequencies and prevents instability. If the delay is several seconds, much as is the case in controlling a vehicle or robotic arm on the moon, then a "move-and-wait" strategy is required, as previously explained in Figure 1.2).

For either the case of a significant pure delay or inertial lag, it can be helpful to have a *predictor display*. This is usually a dynamic model of the controlled process that operates on a much faster timescale than the real system. By inputting the current command to the fast-time model, a display can overlay what is expected to happen after the time delays/lags of the actual system dynamics, and this mitigates or even prevents instability.

For exercising control in a VR environment, there is currently no need to worry about long time delays. However, in older VR systems, the graphics computation time produced a small delay. If the HMD wearer turned her head, and the graphic display was delayed even a small fraction of a second, then that was very disconcerting.

How is the sense of presence affected by the mode of control? When the human operator is "in the control loop", the system being controlled is continuously dependent on operator sensing and action. If the operator sees the objects being telemanipulated relative to the surrounding environment and is continuously adjusting the joystick or yoke control or steering wheel, the human and vehicle are tightly coupled. The operator cannot escape. Hence, the operator by definition is telepresent in the control task and environment. This would be the case especially in flying an aircraft or driving a car while looking out the windscreen. However, if one is flying by instruments, not looking out the window, the sense of presence can be compromised. If the weather is such that the outside scene is in complete fog, and the pilot cannot see the ground, the same is true. Such a compromise on presence likewise applies to the driver of an automobile on autopilot.

Combining computer automation with human control. There are various aspects of controlling a robot where some degree of automation is helpful. For example, when Neil Armstrong first landed the Apollo spacecraft on the moon, he was manually controlling some degrees of freedom (e.g., adjusting forward motion as he was choosing a suitable landing site), but automation was controlling other degrees of freedom (e.g., keeping the orientation of the lunar excursion vehicle upright). These days mixing manual and automatic control is common for many aspects of controlling a robot.

One example is *resolved motion rate control (*Whitney, 1969*)*. Suppose the human wants to control the endpoint motion in six DOF of a multi-link robotic arm. How much to move each of the links to achieve the desired motion? It depends on the particular geometric configuration the arm is in. Trying to control each link manually leads to confusion and catastrophe because the human cannot do the angular geometric calculations in her head, and so the endpoint does not go where she intends it to go. However, a computer can keep track of the geometry and easily perform the calculations, so that with resolved rate control all the human has to do is guide the endpoint in the desired direction.

Another example is where the computer serves as an intermediary to set the stiffness of each joint. Consider the marvels of the human body. If you are catching a baseball, you place the mitt in exactly the right place and keep your arm stiff in accordance with where you think the ball will go. By contrast, if someone tosses you a water-filled balloon, you relax your tendons to make the catch as soft as possible and not burst the balloon. There is a game to see how far away two people can be while tossing a water-filled balloon back and forth. Early industrial robots were all very stiff, and, in the process, they broke many objects they were handling. Some robots now make a point of having stiffness (or its opposite, compliance) under computer control as a function of what the human requests.

There have been many other advances made in enabling greater autonomy in remotely controlled systems. Artificial intelligence (planning, scheduling, and execution), behavior-based, Bayesian networks, causal inference, partially observable Markov decision process, probabilistic robotics, machine learning, etc. applied to robots have enabled many of the advanced capabilities we see today. The subject of artificial intelligence and advanced computer decision processes is too large to address in this book. I would note that some of these advances have also caused some of the shortcomings/concerns that we identify in Chapter 5.

A critical challenge in controlling a telepresence robot is how to navigate through or among obstacles. The obstacles can be rocks on a rough terrain, machinery in a factory, shelves in a supermarket, people on a sidewalk or in an airport terminal, or vehicles on a roadway, for example. The problem of navigation is easier if the objects are static, but if they are people or other vehicles that are moving, there is the problem of predicting where the objects are moving and how to avoid them. Robot navigation is an active research field, and that function can be performed by the low-level computer (much as it is done by low-level neuromuscular computation when one walks through a crowd).

Levels of automation. Either the actual telerobot or the VR avatar can be automated to some extent (degree of local autonomy). What degree or level of automation is most appropriate is *context-dependent*. To help the planner make this determination, assuming there exists some capability to adjust the parameters of the telerobot/avatar, it is worth considering the implications of various levels. In general, the participation of the computer along with the human can be thought of in terms of what has come to be called *levels of automation*. While it is obvious that there are many variables that apply to any measurement, leading to multidimensional scales of level of automation, a one-dimensional scale of levels of automation suggested by Sheridan et al. (1978) has provoked much discussion in the literature of human–machine systems (see e.g., Kaber and Endsley, 2004). One version of the original scale is

Scale of levels of automation

1. The computer offers no assistance, the human must do it all.
2. The computer offers a complete set of action alternatives.
3. The computer narrows the selection down to a few.
4. The computer suggests one.
5. The computer executes that suggestion if the human approves.
6. The computer allows a restricted time to veto before automatic execution.
7. The computer executes automatically, then necessarily informs the human.
8. The computer informs the human after execution only if he asks.
9. The computer informs the human after execution if it, the computer, decides to.
10. The computer decides and acts autonomously, ignoring the human.

Multidimensional scales of automation are easily conjured up, for example, one dimension having to do with sensing capability, another having to do with computational capability, another with action capability, another with degree of interaction/

approval by the human operator, and so on. Clearly, the usefulness of such scales is for planning, thinking hard about what might be most effective. They are in no sense optimization algorithms.

The computer-based adaptive internal model. An important idea that has found its way into modern control thinking, whether taking the form of a human operator's mental model or being instantiated as a physical system, is the so-called *internal model*. It is not a simple error-correction as described above. Very briefly, a best current representation of the state of external reality is input to a policy or *control law*, which incorporates the cost–benefit criteria for what is best to do under all possibilities considered. This then determines overt action on the real physical environment, which in turn determines a new state of the environment. The measure of the new state is compared to the previously measured state, and any small discrepancy is used to adjust the internal model. This approach will be elaborated on later in Chapter 3. It is the same approach that is embodied in computer processing for purposes of automatic control and has also become a model finding favorability in cognitive science as a way of discovering physical reality.

Supervisory telepresence. Many tasks these days are automated, but with humans operating in a *supervisory control* mode (Sheridan, 1992). What is meant by that is that the human intermittently instructs (programs) the machine for parts of the task and then monitors that the machine does what is requested. If the machine complies and the immediate task is accomplished, the human supervisor may then instruct the machine to perform the next step in the process, and so on. Supervisory control is applied when the aircraft pilot programs the autopilot with instructions to fly to a new waypoint or go to a new altitude. Similarly, the so-called self-driving automobile can be programmed to stay in the right lane and drive at a given speed. Such supervisory control was mentioned earlier in describing experiments in my MIT lab when there was a time delay in sending signals to the moon and getting feedback.

Experiencing some degree of presence in the control task environment is critical even though the operation is controlled in supervisory fashion. There are several ways the human supervisor of the aircraft or self-driving car can gain a sense of presence in the task:

1. The aircraft pilot supervisor can continually look out the window at the ground or clouds to gauge orientation though not be active in the control loop. So too, the driver of the self-driven car can continuously look out the window and maintain awareness of any hazard in the road or with other vehicles. (The manufacturer insists that this is required!)
2. The pilot can continually monitor a computer-generated map or other pictorial display that shows the continuously changing position of the aircraft relative to the ground terrain, other aircraft, or threatening weather cells. Similarly, the driver can maintain his position of a moving map (similar to many navigation systems).
3. The out-the-window and the computer-generated views can be combined using augmented reality to combine images. For example, if the ambient

weather is very foggy, and the aircraft is coming in for a landing, an image of the runway can be projected on the windscreen. If the fog sufficiently clears as the aircraft approaches the ground, the pilot may use the out-the-window view to land. If the fog does not clear, the pilot can execute a go-around or else put the aircraft into an auto-land automatic mode. A similar augmented reality approach is viable in the case of the automobile, though as of this writing it has not been implemented.

Up to this point, we have considered only a singular dimension of continuous control for aircraft and automobiles. To broaden the perspective, consider a human supervisor in a chemical plant, a nuclear power station, or a large manufacturing plant. The human supervisor is intermittently attending to many different control tasks, which may involve pushing buttons, adjusting levels and flows, or calling for computer programs to execute. Each subtask may be located in a different part of a building or complex of equipment The supervisor is allocating attention among these many subtasks, as shown in the bottom part of Figure 2.16. At the lower level are one or more computers exercising local control on particular subsystems (sometimes called *local autonomy*).

At the top of Figure 2.16 are shown the many decision categories the supervisor of such a complex system must attend to. They include the following:

1. **Planning** includes having some mental and possibly explicit graphical model of the whole system and how the parts interact. This also includes having some notion of the trading relations, or compromises in the use of time, energy, and other resources that must be allocated, and finally formulating some management strategy.
2. **Teaching** (programming) the task elements according to the plan.
3. **Monitoring** the automatic execution, which includes allocation of attention to each subtask, estimating the state of progress, and diagnosis of any abnormality or unanticipated events.
4. **Intervening** in the automation to modify the plan as necessary.
5. **Learning** from the whole experience so as to improve the supervisory process.

In contrast to supervising the continuous control of air or highway vehicles, the supervision of a stationary chemical, nuclear, or manufacturing plant poses a very different telepresence challenge. Such facilities are spread out in space, with equipment

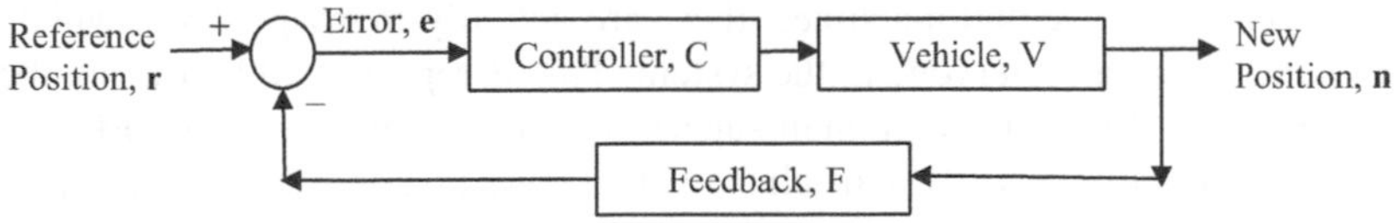

FIGURE 2.15 Feedback control.

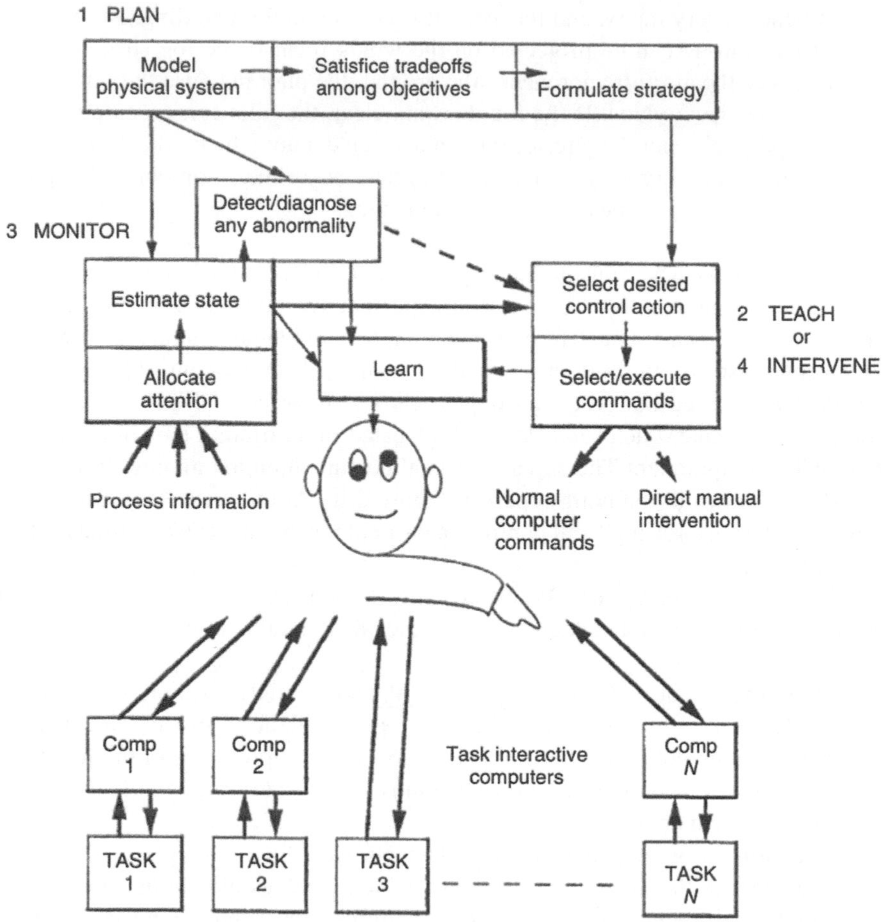

FIGURE 2.16 Generalized human supervisor.

that from time to time must be inspected or adjusted. Although much of the control can be done remotely, e.g., from a control room, not all can. The situation requires that some person or mobile robot make the rounds to check for any problems at various locations. At times, it is hazardous to go into some areas. So a telepresence robot is very useful, assuming it has sufficient mobility and can get close to what needs to be inspected or can manipulate controls that are remotely located.

In modern factories, hospitals, air traffic control systems, or other complex arrangements of people and machines, there are actually many points at which the human supervisor can intervene in the system. This is represented by the numbered arrows in Figure 2.17: (1) The human supervisor can modify the trade-offs among values in the control law; (2) can change the physical control interface; (3) can change the goal or reference input for the whole system; (4) can invoke test actions on a simulator that is run in fast time (which uses the reference input previously set);

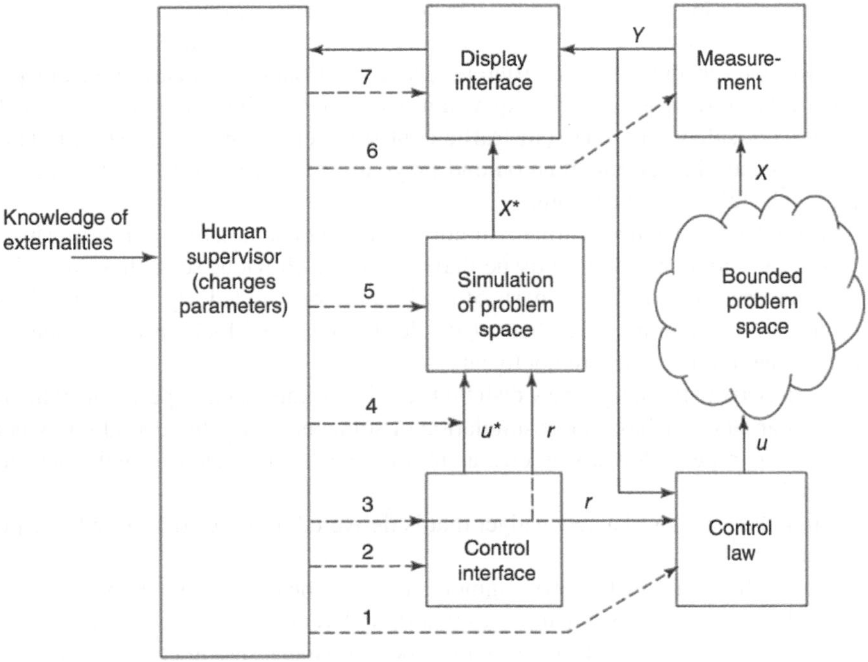

FIGURE 2.17 Diagram showing seven types of parameter adjustments the human supervisor can make.

(5) can modify the parameters of the simulator; (6) can change the way system state is measured; and (7) can modify the display interface. Think of this diagram as a complement to previous Figure 2.16, where a busy supervisor can interact at various points in a complex system, robotic, or otherwise (Sheridan, 2011).

Future system supervision will involve telepresence simulators in various roles, using VR or not. Internet YouTube recordings try to help, but I have found them to be frustrating because the presenters typically talk too fast, point to details in the interface that are not obvious, and act as though understanding the technology and the tools of operation are obvious to the user. Bringing the user/observer into a VR simulation is a big challenge. The user should be allowed to set her own pace and allowed to try things out, point to features, and ask questions like "What is this?, How do I … etc." in the process of being introduced to the technology. Modern question–answering technology should be capable of answering most of the questions.

Imagine a human operator of a nuclear power plant. At some stage of the operation, she feels that a particular valve should be opened. The safety procedures suggest that, but she has some doubts as to what the implications are for opening that valve. She asks the computer to run a simulation of the effects of opening the valve, observes what would happen, and feels better that what will happen is what she intended.

Consider an air traffic controller who is shepherding aircraft into a particular airport. Because of bad weather one aircraft is delayed, and the question is whether to modify the schedule to put a different aircraft in front of the delayed one, or just delay all the arriving aircraft. The question is whether retaining the original schedule can work safely. The ATC (air traffic control) operator invokes a VR simulator which picks up the current situation and plays it out, suggesting that retaining the current schedule would be unsafe.

Consider a surgeon performing a tumor operation on a human brain. It is critical that incisions not damage more tissue than necessary. Previous scans have revealed where the tumor lies relative to parts of the brain to be avoided. By using an HMD, the surgeon can in effect place herself inside the brain and look around to see the extent of the tumor and where not to cut.

A housewife plans to try a new dish that requires some special operations that are not familiar. She watches a short simulation of someone doing that operation, which she can stop, repeat, slow down, etc., as many times as necessary to make her feel comfortable.

A man has purchased a new gadget that is delivered in pieces that have to be put together.

By using an HMD that allows augmented reality, he can see which pieces lying on the table correspond to which pieces in a short VR demonstration, and have to be assembled in what order, making the whole operation much smoother than otherwise.

Digital twins. A *Time* magazine article (Chow, 2022) describes a number of examples of such 3D-assistive simulation software and computer graphics using the term "digital twins". Digital twins are computer simulations that are continually updated to match some real-world system. Sensors plugged into the real system allow the updating of the digital twin.

- A BMW plant in Regensburg, Germany, operated a digital twin that duplicates (models) machinery, materials, movements of workers (avatars), etc., and how they all interact.
- NASA uses replicas of spacecraft and all their software/hardware functions that operate in real time during development, testing, and deployment of the vehicles.
- Architecture firms build digital twins of the buildings they are designing so that clients can observe and, wearing HMDs, can walk through and experience the buildings before they are built. Actually, this was one of the first applications of VR that this author experienced as his introduction to the technology.
- The medical community is using digital twins during planning and execution of complex medical diagnoses.
- Vehicle developers are using digital twins during crash tests to better understand safety aspects.

Nuclear power companies, community developers, transportation planners, and many others are making use of such computer simulations.

Many software firms are offering 3D simulation software, such as Nvidia's Omniverse. Now questions of fidelity, privacy, and cybersecurity are being raised. (Regulatory issues will be discussed in the last chapter on Challenges).

TELEPRESENCE COMMUNICATIONS

The internet of things. The internet of things (IoT) is a network of connected smart devices, including robots, that is gradually emerging globally. But it can also be a security nightmare. IoT can bring the same efficiencies to physical manufacturing and distribution that the internet has long delivered for knowledge work. Millions, if not billions, of embedded internet-enabled sensors worldwide are currently providing an incredibly rich set of data that companies can use to gather data about their safety of their operations, track assets, and reduce manual processes. Researchers can also use the IoT to gather data about people's preferences and behavior, though that can have serious implications for privacy and security. There are more than 50 billion IoT devices as of 2020, and those devices will generate 4.4 ZB of data this year [a zettabyte (ZB) is a trillion gigabytes]. (Fruhlinger, 2020). Vermesan et al. (2020) provide a detailed review of the communication and control of "things" over the internet, including many sub-internets in different application areas.

Deep space communication. Spacecraft send information and pictures back to Earth using the NASA Deep Space Network (DSN), a collection of big radio antennas. The antennas also receive details about where the spacecraft are located, and how they are doing healthwise. NASA also uses the DSN to send lists of instructions to the spacecraft. Ground segment facilities located in the United States, Spain, and Australia support NASA's interplanetary spacecraft missions. For example, Deep Space Station (DSS) 43, the 70 m antenna at the Canberra Deep Space Communications Complex, has a K-band radio astronomy system covering a 10-GHz bandwidth at 17–27 GHz (https://spaceplace.nasa.gov/dsn-antennas/en/).

REFERENCES

Caeiro-Rodríguez M, Otero-González I, Mikic-Fonte F, et al.: A systematic review of commercial smart gloves: Current status and applications, *Sensors* 21(8): 2667, 2021. https://doi.org/10.3390/s21082667.

Chow A: How digital twins are transforming manufacturing, medicine and more, *Time Magazine*, January 24, 2022.

Fruhlinger J: http://www.networkworld.com/article/3207535/what-is-iot-the-internet-of-things-explained, 2020.

Kaber D and Endsley M: The effects of level of automation and adaptive automation on human performance, situation awareness and workload in a dynamic control task, *Theoretical Issues in Ergonomics Science* 5(2): 113–153, 2004.

Landsberger S and Sheridan T: *Parallel Link Manipulators.* US Patent 4666362, May 1987.

National geographic explore VR is a game available on M eta's oculus quest, 2020.

Sheridan T: Adaptive automation, level of automation, allocation authority, supervisory control and adaptive control: Distinctions and modes of adaptation, *IEEE Transactions on Systems, Man, and Cybernetics* 41(4): 662–667, 2011.

Sheridan T: *Opto-Mechanical Touch Sensor*, US Patent 4599908, July 1986.

Sheridan T: *Telerobotics, Automation and Human Supervisory Control*, Cambridge, MA, MIT Press, 1992.

Sheridan T and Ferrell R: *Man-Machine Systems*, Cambridge, MA, MIT Press, 1974.

Sheridan T, Verplank W and Brooks T: *Human and Computer Control of Undersea Teleoperators*. Report of the MIT Man-Machine System Laboratory, December 1978.

Vermesan O, Bahr R, Otella M, et al.: Internet of robotic things intelligent connectivity and platforms, *Frontiers of Robotics. AI*, September 25, 2020. https://doi.org/10.3389/frobt.2020.00104.

Wenzel E: Localization using non-individualized head-related transfer functions, *Journal of the Acoustical Society of America* 94: 111, 1993.

Whitney D: Resolved motion rate control of manipulators and prostheses, *IEEE Transactions on Man-Machine Systems* 10(2: 47–53, June 1969.

3 What Is Telepresence? What Is Reality?

In this chapter, we first consider what variables are key to the sense of telepresence of both the robotic and virtual varieties. Then we look at various proposals for how to measure presence, some using subjective scales and some based on objective metrics. Finally, the chapter reviews a number of philosophical perspectives on reality that have been motivated by the new technologies of VR and robotic telepresence.

Attributes of telepresence. Sheridan (1992a) proposed that there are three principal attributes of "telepresence":

1) The first is the **extent of sensory information**. In terms of vision, this implies the frame rate of the display, the resolution of the display, and the information in terms of color and grayscale per pixel (see Figure 3.1). The frame rate is quantified as frames per second, the resolution is quantified as pixels per frame, and the information depth is quantified as bits per pixel of color or grayscale differentiation. Thus, we have the product of these three terms being bits per second. Information experiments in my lab clearly showed that compromising any of these would compromise the display and hence reduce a sense of presence. But the effect is very dependent on the task context. For example, in reading an X-ray (or examining any image where the frame rate is largely irrelevant), only pixels per frame and bits per pixel make the difference. In a dynamic task such as an umpire predicting whether a thrown baseball is a ball or strike, frames per second and pixels per frame are critical but color and grayscale are not. The corresponding variables apply to hearing: the spatial location resolution is analogous to pixels per frame, the frequency resolution is analogous to frames per second, and the intensity resolution is analogous to bits per pixel. Similar spatial, dynamic, and intensity variables apply to haptic and tactile displays.
2) A second major attribute of presence is **operator control of sensors**. That is manifest in the operator being able to turn her head and adjust her gaze to arbitrary positions. It is noted that in early VR systems, there were time delays between when the operator turned her head and when the corresponding visual display would catch up to displaying the image that would be seen from that location and angle. That was extremely disconcerting, but with faster VR computation that problem has mostly gone away. In long-distance robotic telepresence, signal transmission time delays cannot be eliminated, so the human operator must move-and-wait for feedback (as described in Figure 1.2). Operator control and corresponding timely feedback for haptic and tactile sensors are similarly critical to the sense of telepresence.

DOI: 10.1201/9781003297758-4

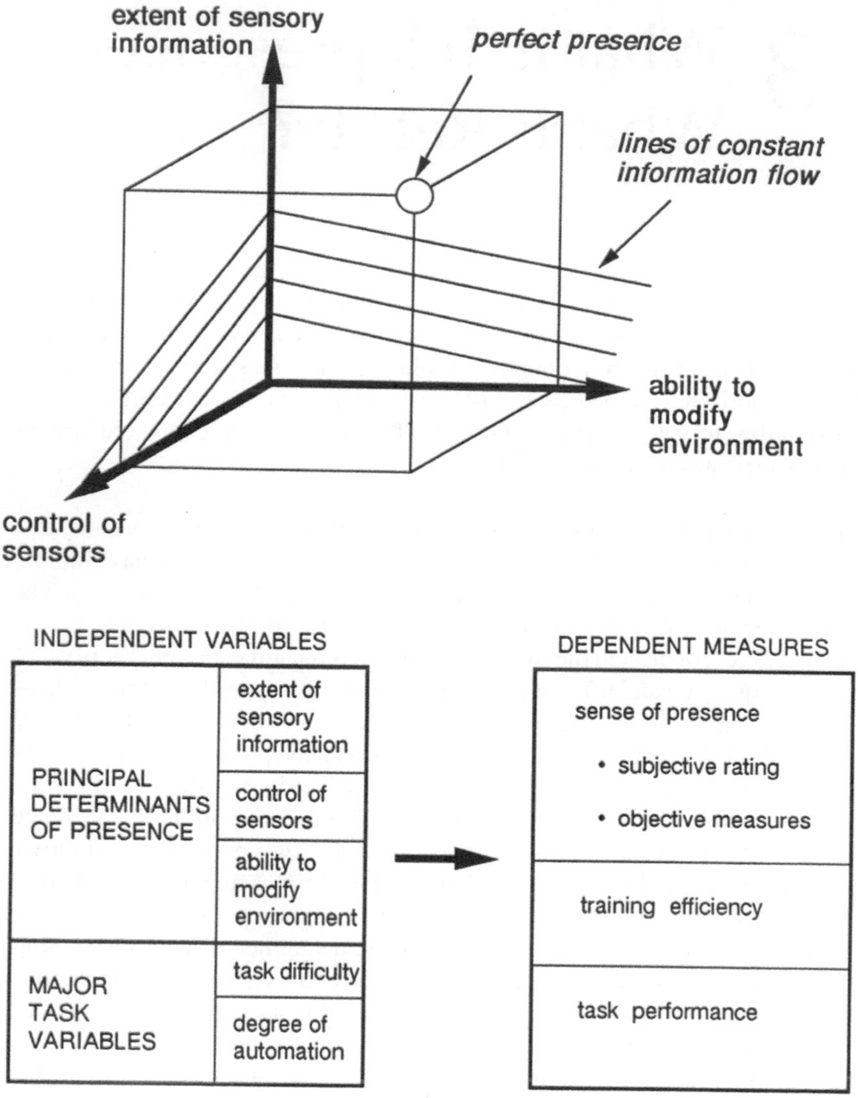

FIGURE 3.1 Factors that comprise the extent of sensory information. In the graph above, the three components can be considered a three-dimensional function. In the table below, the independent and dependent variables are indicated.

3) A third critical attribute is the **ability of the human operator to perform some mechanical task** in the remote or virtual environment. To have accurate and timely visual auditory, haptic, and sensory stimulation, *but be unable to make anything happen* at the remote site, simply kills the sense of normal telepresence. To have the environment refuse to respond to efforts to manipulate it makes evident that it is not real.

There are other factors that affect the sense of telepresence. Table 3.1 provides what variables are featured in research by several different authors.

MEASURES OF TELEPRESENCE

Subjective measures of telepresence. Slater et al. (1994) used three questions (actually statements on which the subject rated agreement or disagreement, embedded in a longer questionnaire) to elicit a measure of the "depth of presence" in a VR environment:

1) *In the computer-generated world I had a sense of "being there"* (7-level scale from "not at all" to "very much").
2) *There were times during the experience when the computer-generated world became more real or present for me compared to the real world* (7-level scale from "at no time" to "almost all of the time").
3) *When you think back on your experience, do you think of it more as something that you saw, or more as somewhere you visited* (7-level scale from "something that I saw" to "somewhere that I visited").

Other questions related to the sense of nausea, prior experience of vertigo, prior use of computers, etc.

Witmer and Singer (1998) posed to subjects a number of questions associated with presence. I repeat them here to provide a sense of how varied such questions can be:

1. How much were you able to control events?
2. How responsive was the environment to actions that you initiated (or performed)?
3. How naturally did your interactions with the environment seem?
4. How completely were all of your senses engaged?
5. How much did the visual aspects of the environment involve you?
6. How much did the auditory aspects of the environment involve you?
7. How natural was the mechanism that controlled movement through the environment?
8. How aware were you of events occurring in the real world around you?
9. How aware were you of your display and control devices?
10. How compelling was your sense of objects moving through space?
11. How inconsistent or disconnected was the information coming from your various senses?
12. How much did your experiences in the virtual environment seem consistent with your real-world experiences?
13. Were you able to anticipate what would happen next in response to the actions that you performed?
14. How completely were you able to actively survey or search the environment using vision?
15. How well could you identify sounds?

TABLE 3.1
Central Features of Technological Features of Telepresence (after Draper et al., 1998)

Approach	Nature of Telepresence.	Causes	Relation to Performance relationrelationship to Performance
Akin et al (1983)	A feeling of actual presence at the worksite	1. Manipulator dexterity 2. Feedback scope and fidelity	Telepresence improves performance
Sheridan (1992a, 1992b, 1996)	User feels physically present at the remote site; compelling illusion; subjective sensation	1. Sensory fidelity 2. Sensory control 3. Manipulability	Telepresence improves performance
Steuer (1992)	The sense of being in an environment; the experience of presence by means of a communication medium	1. Vividness 2. Interactivity	Telepresence improves performance
Zeltzer (1992)	Sense of being in and of the world	1. Autonomy 2. Interaction 3. Presence	Telepresence improves performance but might make tasks more difficult and fatiguing
Slater and Usoh (1993); Slater et al (1994)	The suspension of disbelief that they are in a world other than where their real bodies are	1. External factors 2. Internal factors	Telepresence improves performance
Witmer and Singer (1994)	Subjective experience of being in one place when physically in another; subjective sensation much like "mental workload"; a mental manifestation	1. Control factors 2. Sensory factors 3. Distraction factors 4. Realism factors	No clear relationship
Schloerb (1995)	The person perceives that he or she is physically present in a computer-mediated environment	1. Information flow 2. Ability to manipulate computer-mediated environment	Performance must reach some minimum level before telepresence can occur
Muhlbach et al (1995)	Sense of sharing space with distant interlocutors	1. Spatial presence 2. Communicative presence	Telepresence improves performance

16. How well could you localize sounds?
17. How well could you actively survey or search the virtual environment using touch?
18. How compelling was your sense of moving around inside the virtual environment?
19. How closely were you able to examine objects?
20. How well could you examine objects from multiple viewpoints?
21. How well could you move or manipulate objects in the virtual environment?
22. To what degree did you feel confused or disoriented at the beginning of breaks or at the end of the experimental session?
23. How involved were you in the virtual environment experience?
24. How distracting was the control mechanism?
25. How much delay did you experience between your actions and expected outcomes?
26. How quickly did you adjust to the virtual environment experience?
27. How proficient in moving and interacting with the virtual environment did you feel at the end of the experience?
28. How much did the visual display quality interfere or distract you from performing assigned tasks or required activities?

Quantitative measures of presence. Schloerb (1995) suggested simple quantitative measures of telepresence, both of which he termed "objective telepresence" and "subjective telepresence".

"The degree of (objective) telepresence is equal to the probability of successfully completing a specified task. The degree of subjective telepresence is equal to the probability that a human operator perceives that he or she is physically present in a given remote environment".

Bracken et al. (2014) proposed the use of secondary task reaction time. That means that experimental subjects are asked to attend not only to the primary task, in this case consisting of a virtual or a telerobotic environment, but also to a secondary task such as a selective choice response to a randomly occurring question. This requires divided attention, and it is assumed that to the degree that the subject performs worse on the secondary task (longer reaction time) the primary task is more compelling, the subject is more "immersed" in it.

WHAT IS REALITY? PHILOSOPHICAL PERSPECTIVES

The advent of virtual and augmented reality, the explosive growth of online gaming, and the huge recent investments of Meta (formerly Facebook), Microsoft, Apple, and Google in these areas have provoked many interesting philosophical questions about reality. What is real? What is meant by reality? Professional philosophers of course have struggled with that question since and before the ancient Greeks. But the fact that technology can now "immerse" an individual in a different reality than the everyday actual experience has begged the question in recent philosophical discussion.

We start with the early (1992) pages of *Presence: Teleoperators and Virtual Environments* (currently called *Presence: Virtual and Augmented Reality*), a publication that the author had a hand in founding. It contained a lively discussion on the ontology of "presence", namely the philosophical aspects of the experience of virtual and "real" reality and being (Barfield et al, 1990: Heeter, 1992; Held and Durlach, 1992; Sheridan, 1992a, 1996; Zeltzer, 1992; Slater and Usoh, 1993; Steuer, 1992; Slater et al., 1994; Schloerb, 1995; Hendrix and Barfield, 1995; Mühlbach et al, 1995; Flach and Holdren, 1998; Witmer and Singer, 1994; Zahorik and Jenison, 1998; Mantovani and Riva, 1999). This topic is obviously critical to virtual reality (VR) research pursuits because the philosophical perspective affects what we do in our scientific and engineering pursuits and how we interpret and describe what we find (Dennett, 1978).

Most authors have assumed that actual and virtual presence are quite distinguishable, that mostly we know where we really are, unless we are asleep, drugged, or just not paying attention. However, by some effort to "suppress disbelief", anyone can experience a sense of presence in everyday activities. For example, a virtual telepresence experience of "being there" comes by way of being caught up in reading or being told a story, watching actors in a play or film, or participating in the same, or interacting with a computer—visually with computer graphics, auditorily with computer-generated sound, or haptically with computer-generated mechanical forces on our bodies.

Some philosophers have asserted that in a deep sense we are always in a virtual world, and that we cannot know reality, that our fallible senses and brains deceive us. They assert that reality is not something fixed, but that we are constantly changing our reality through our actions.

Numerous proposals claim to be at odds with the conventional Cartesian perspective that has formed the basis for science and engineering for so long. Social science—in contrast to the so-called "hard" sciences—has seen a variety of paradigms enter the intellectual fray, grow in popularity and fashion, and then fade away. My own pedestrian perspective is that "presence" is tolerant of many of these supposedly conflicting ontologies, and seems to live happily in a Cartesian world that has already proven its mettle in various hard science and engineering applications.

Not knowing how best to organize a chapter on philosophical aspects of telepresence and virtual reality, I hope it will be acceptable to first review some classical ontologies of reality. Then I will suggest an idea that has emerged in engineering control theory that I believe is useful in the broader philosophical context. Finally, I will review various ideas about reality from some current philosophers.

Descartes. The French philosopher Rene Descartes postulated a dualism between *res extensa*, objects located outside the mind, and *res cogitans*, objects located within the mind. The Cartesian view—that there are clearly separable mental and physical domains, what Zahorik and Jenison (1998) refer to as the *rationalistic tradition* and what Manovani and Riva (1999) call *ingenuous realism*, pervades Western thinking today, including cognitive science as well as physical science and engineering. There are objective measures and subjective measures, and we know which is which.

Most of today's research into virtual reality and how it differs from "actual" reality adheres to the rationalist tradition.

Heidegger. The German philosopher Martin Heidegger (1927) was absorbed with the question of *being*. He rejected the Cartesian view of a dualistic reality and asserted that all meaning, hence all reality, is conditioned by interpretation, including the beliefs, language, and practices of the interpreter. According to Heidegger (Zahorik and Jenison, 1998), we are *thrown* into situations where an action is unavoidable (*throwness* in Heidegger terminology after translation from German), the result of such action is unpredictable, and a stable representation of the situation is impossible. In normal use of a tool or other object (e.g., in hammering), the tool becomes transparent to the user, who cannot then conceive of the tool independently (it is *ready-to-hand* in Heidegger-speak). However, if some abnormality occurs (e.g., the hammer slips) there is *breakdown*, and the tool then can be conceived in the mind (it becomes *present-at-hand*). Normal *being*, in the Heidegger view, means complete involvement in a dynamic interaction—in which subject and object are not separable—and only by stepping back and disconnecting from that involvement can a person perceive the elements of the situation. Does the perceived reality of telepresence and virtual reality conflate with Heidegger's *being*?

J. J. Gibson. The American psychologist J. J. Gibson has carried the Heidegger view forward, especially with respect to human perception (Gibson, 1979). According to Gibson, perception is the acquisition of information that supports action, especially with regard to constraints on action (Gibson calls these constraints *affordances*). Actions affect the environment, and the environment in turn affects the action in complete reciprocity. Perceptions are true, in the Gibson view, to the extent that they support action in the environment. As Zahorik and Jenison (1998) state in comparing Heidegger and Gibson to the rationalistic tradition, "ecology replaces phenomenology, and presence is tantamount to successfully supported action in the environment".

The rationalist tradition and the Heidegger–Gibson perspectives have been touched upon briefly because they have been posed by others (especially in the journal *Presence*) as being opposing and incompatible with Descartes. The gist of the incompatibility is in whether perception is different from and independent of action, i.e., the two have different dimensions (the rationalist perspective)—or whether, as Flach (1999) has put it with respect to Heidegger and Gibson, perception and action have the same dimensions, with the perception vector incorporating action constraints and the action vector incorporating perceptual constraints. Presumably, the action constraints are directly available to the nervous system as it senses and perceives the physical world. "It does not have to translate from one coordinate system (sensory) to another (motor)."

Of course, there are many other ontological perspectives with corresponding incompatibilities. As pointed out by Flach and Holdren (1998), Newton's use of physical space and time as the absolutes of reality contrasts with Einstein's merger of space and time. Fechner's introduction of subjective scales mapped against physical scales and Helmholz's distinction between distal (physical) and proximal (sensory) both seem to lie within the rationalistic tradition, while Dewey—in emphasizing coordination between sensing and acting and the environment—tilts in the Heidegger–Gibson direction. Kant (1781), on the other hand (in *Critique of Pure Reason*), would seem more consistent with Descartes:

> But the conjunction of representations into a conception is not to be found in objects themselves, nor can it be borrowed from them and taken up into the understanding by perception, but is on the contrary an operation of the understanding itself, which is nothing more than the faculty of conjoining a priori and of bringing the variety of given representations under the unity of apperception.

A current trend in cognitive psychology is "naturalistic decision-making" (Zsambok and Klein, 1997), which claims to reject traditional normative decision theory in favor of the Heidegger–Gibson approach.

From this sample, one might assert that the alleged incompatibility is simply a matter of history, that the older rationalist tradition is simply being replaced by a newer paradigm that is more along the lines of Heidegger–Gibson. Or is it? One might point to modern neurobiology. (see, for example, Edelman (1992) or Crick (1994). The question is whether the evidence for sensory-motor mapping fits better with the rationalist tradition or the Heidegger–Gibson tradition. And how much is in the head and how much is "out there"? And where does this leave "reality" and "presence"?

Divine presence. While hard questioning about divine presence is not particularly popular these days in the scientific literature (indeed, it risks alienating the reader!), there is no dearth of it in the popular literature. But why bring divine presence into this discussion? Precisely because belief in God represents an ultimate ontological challenge with regard to perceiving and interacting with "reality" and "presence", and (depending upon one's belief) with respect to accepting virtual reality as objective reality, or even making the distinction. While God is apparently not Himself (Herself, Itself) arguing a philosophical viewpoint on reality as are the other three, the sacred texts (Bible, Koran, etc.) are taken by believers as arguing on behalf of God's presence and reality.

To the traditional believer, God is the ultimate reality, knowable through scriptural interpretation (hermeneutics again!) and "present" everywhere, even though presence has no sensory manifestation in the way that operational physics would demand (repeatable observation by different people). For the latter reason, God as manifest to the believer can be said to be a reality whose "presence" is enhanced by faith and active participation in religious exercises (for example, communal prayer, singing, etc.). Ontologically, God may be said to be a virtual reality (as contrasted to physical sensible reality), whose "presence" is enhanced by suppression of disbelief through the above-mentioned participation, in the same way that immersion in a computer-graphic virtual environment is enhanced.

To many others, God is a metaphor for the universe of what is unknown and unknowable (Sheridan, 2014). In this case, active suppression of disbelief can enable one to conjure up spiritual inspiration and wonder at the extent of that unknown universe. In either case, the sense of divine or otherwise spiritual presence or reality is enhanced by voluntary effort on the part of the observer. The difference is that the criteria of the believer for accepting an internal model of reality (with respect to God) differ from those of the nonbeliever.

Both believer and nonbeliever admit to a distinction between the physically sensible and the spiritual "realities", and that would seem to fit the rationalist tradition. Both would have to accept that Heidegger's "ready-to-hand" phenomenon characterizes being "thrown" into religious or otherwise spiritual experience, while some interruption (say, a fire alarm) can result in "breakdown" and a "present-at-hand" experience. In both cases, one would have to agree with Gibson that action is what makes a difference in the perceived reality: worshipful action affects the perception of God, and perception of God has in it the affordance of interaction. But how much of this is "in the head" (and therefore, by definition, virtual) and how much is "out there" and therefore real?

Model-based discovery of "reality". Having swung from secular philosophy to divinity (at least spirituality), let us now swing in the other direction to engineering. In particular, let's examine briefly an information-processing technique—*estimation theory*—that has proven very powerful for observing and perceiving physical reality (Sheridan, 1992b). This description will be qualitative and provide only the gist of the idea.

Suppose an animal or machine could sense its external environment perfectly and continuously in time (through some hypothetical suite of sensors, it matters not for this argument), but was unable to relate what it sensed to any internal model of the environment, any map, and any procedure. It would be reduced to an error-nulling controller as previously described. It could be programmed (or motivated) with a set point, and then it could take corrective action in proportion to its perceived instantaneous error between the set point and the perceived current position or state. This, of course, is the way many control devices do work, but obviously we could not call this a very intelligent controller. The reference set point may quickly become obsolete.

Suppose, at another extreme, an animal or machine had no means to measure the external environment, but it had a perfect *internal model* of a *fixed environment* at some initial point in time. Suppose also that it had perfect interoceptors (internal sensors), so that it could measure and remember every detail of the position and force trajectories of its own action. By integration of these actions over time and comparison to the fixed environment, the animal or machine could "know" exactly where it was with respect to its environment, and then act accordingly. This would be perfect at so-called "dead-reckoning". But if something in the environment changed, the model would no longer be valid.

Finally, suppose that exteroceptors, interoceptors, and model were all present, but were imperfect. What then? Is there some way to get the best from both error feedback and dead-reckoning combined? This is surely the reality with real animals. And this is where estimation theory comes in. In engineering, it is currently considered the best way to discover the momentary environment—by acting on it or in it.

Figure 3.2 shows the diagram of the process of estimation, as commonly considered by control engineers and estimation theorists. The boxes represent

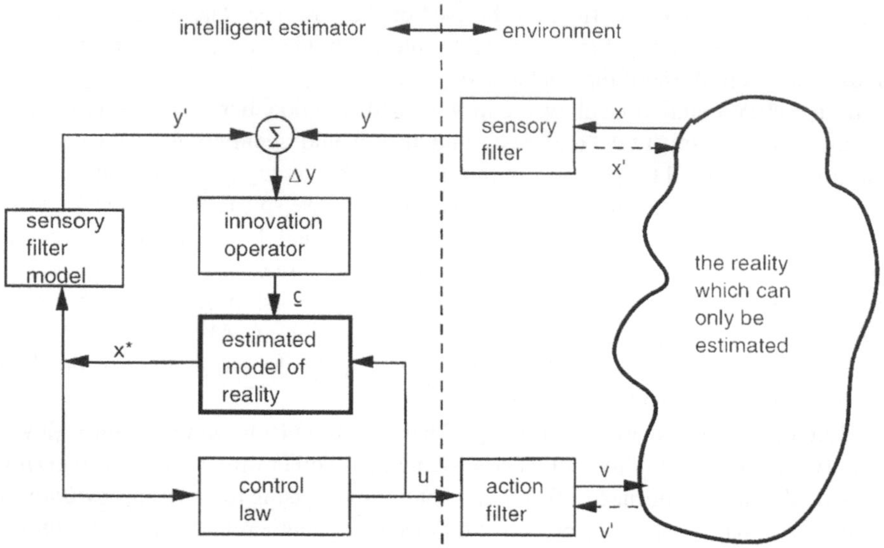

FIGURE 3.2 The process of estimation.

operations that in various applications are well defined and usually mathematical, typically linear differential equation transfer functions. To the left of the dashed line is the intelligent estimator, the animal, or computer, and to the right is its environment. The blob on the right represents the true reality "out there". We assert that this true reality can never be known, but can be only estimated because of noisy "filters" that lie between the information processing internal to the animal or machine and the outside world. There is a sensory filter on the input (afferent) side and an action filter on the output (efferent) side, and these filters pose distortions such as random noise, nonlinearities, non-stationarities, and time delay—and in effect limit observation to but a minute portion of the potential spectrum of variables that are "out there". The sensory and action filters represent the well-experienced constraints on the animal's own sense organs and limbs, or the equivalent for the computer, i.e., limited robotic sensors and mechanisms for manipulation or locomotion.

The arrows represent causality or signal flow, and their lowercase letter labels are the key variables. In general, all the variables are vectors (quite large ones). For qualitative characterization purposes here, we need not be concerned with their size. At the upper right, x is the *state* of reality, the unknown that we seek to know (and therefore to estimate). It is sensed by a constraining sensory filter that is represented (to the extent that it is known from experience) in the biological or electronic brain on the left of the dashed line. The covariable x' always exists when two systems interact. In physical interactions, for example, if x were voltage measurement, x' would be the current drawn from the environment by the voltmeter; if x were mechanical position, x' would be the force imposed by the strain gauge or other device. Mostly,

x' is a small effect, but one can never measure something without affecting it (the Heisenberg principle?). In social interactions, x' is the effect the observer has on the observed.

On the left side, x^* is the sought-after estimate of x. (We will say in a moment how it is derived.) Initially, on the basis of a best estimate of the state of reality x^*, the animal or machine decides what to do about it (a *control law*), generating a command to the action filter (muscles or motors), the output of which is v, a mechanical force or position. (Both animal or robot actions are usually mechanical actions.) The covariable for v is v', and v is the mechanical position or force complement to v', where the product of v and v' is energy. In this case, the action on the environment generates an action back on the muscles or motors: the intelligent entity is changed in the process of making a change, this time (at least in a physical sense) to a more significant extent than with sensing, because the energy level with muscles is greater than that of sense organs.

The heavy-lined box on the left is a *model of reality* (including a model of the action filter), which is driven by u, which similarly drives the action filter and true reality. The output from the model of reality is the state (of reality) estimate x^*, which can be seen to be an input to the sensory filter model as well as the control law. Although x^* was initially posited as a best guess, it is gradually improved, based on a discrepancy between the output y' of the sensory filter model and the output y from the actual sensory filter. This discrepancy Δy is the input to an *innovation operator* that determines how the model is adjusted, driven into conformity with sensed reality (by c, the innovations change variable). Thus, the whole system is set up for the internal model of reality to continually work to mimic (discover, estimate) the state of the true reality.

Notice that the estimator works on the basis of feedback control, not the simple and well-known form of feedback control we described in Chapter 2, but rather a scheme that bootstraps a model into progressively closer matching how the measured environment works.

To make an analogy with human behavior, consider that at times visual or kinesthetic feedback is noisy (e.g., rapidly walking up or down stairs), such that in no way could an error sense guide the feet to move and land properly on each step. Instead, the nervous system's internal model happily operates "open loop" (dead-reckoning) for these short intervals when simple feedback is not adequate. But in the long run, the body depends on visual feedback to see where it is going and not collide with things. This is essentially what control by model-based estimation is doing.

This estimation scheme forms the basis of modern control theory, and it is sometimes called an *observer* or *Kalman estimator* (Kalman, 1960). In modern (or "optimal") control, there is a related theory (which need not concern us here) for shaping the control law, given x^*, in order to optimize with respect to a given objective function.

Figure 3.2 can also be said to characterize a predictor instrument, which is used, for example, when telemanipulating from earth a robot in space, where there is a communication time delay in both the action filter and the sensory filter (Sheridan, 1993). The model in this case incorporates the expected kinematics/dynamics of the robot manipulator, but not the action delay. A computer-graphic virtual image (x^*) of the robot manipulator leads in time the true reality of robot manipulator action in space to the extent that the round-trip communications time delay is excluded from its model. The human operator on earth can then control directly from this virtual representation, since, by circumventing the delay, an instability in control can be ameliorated (since some feedback is in phase with error and thus would grow with time). The human operator also observes the video picture y coming back from space (on which the graphical model image is superposed), and insofar as it does not follow, x^* can modify bias, gain, or other model parameters to get a better registration of the model with the video. In this way, the operator can gain satisfaction that his actions are having the desired effect in the remote space environment.

Getting now to the reason for this brief foray into engineering, the estimation paradigm suggests the following about ontology:

1. "Real" reality can never be known but, because of sensory and action constraints, only estimated.
2. This estimate takes the form of an internal action-sensation-model-refinement, in which representation of the environment can be evolved through action upon the environment, sensing the result, and refinement of the model to get a better prediction.
3. The action and sensing are not random, but are conditioned by the action and sensory filters. For this reason, and because of time limitations, the model will fit only that subset of reality that has been experienced.
4. The state vector of the internal model will include both action and sensory dimensions.
5. Action upon the environment (energy exchanges through the action filter) can change the environment, especially in the neighborhood of the action. Measurement (through the sensory filter) can also affect the environment, but usually less so.

An eclectic ontology of presence and reality. The rationalistic tradition asserts that the subjective mental reality is different from the objective physical reality. The Heidegger position asserts that subjective and objective experience cannot be separated from one another because, in normal behavior, a person interacts dynamically with the environment and there is reciprocity of effects. Only by disconnecting with reality to some extent (breakdown) can perception of the difference occur. The Gibson position emphasizes the primacy of action with respect to environmental constraints (affordances) and further emphasizes the reciprocity of interactive effects.

But is the rationalist position really at odds with the Heidegger–Gibson position? Is distinguishing subjective or mental reality from environmental reality incompatible with the idea that human and environment interact and affect one another? I do not believe so—if one is especially careful of the meanings of some terms such as *subjective* and *objective*. I believe that the methodology of estimation theory provides a paradigm for accommodation.

In the estimation paradigm, objective reality is what one converges to in a sufficiently stable environment, using scientific operational methods (repeatable experiments with different observers), and with many measurements by proven instrumentation (well-known and trustworthy action filter and sensory filter). But that reality is never absolute, only the limiting case of what the estimation model produces (the documented aggregate of mental or subjective perception and cognition).

Given a careful and responsible estimation of objective reality in some well-defined domain, it is not too much of a stretch to call the resulting model "objective" and use that as a yardstick in other experiments. That, essentially, is what we do in our perceptual psychology and virtual reality experiments every day.

In the worlds of science and engineering, the reciprocity of effects between the actor and the environment is well established. The Gibson "affordance" is commonly known in physics as *impedance*, and is typically expressed in differential equation form representing the laws of Newton, Coulomb, Ohm, Faraday, or others, depending on the particular mix of physical media that connect across some interface. In fact, special bond-graph methods have been devised for deriving these equations directly from the topology of physical elements involved (Karnopp and Rosenberg, 1968). Particular impedance characteristics can be identified at interfaces between active machines and passive physical environments, and the same is coming to be true at interfaces of the human body with physical environments.

While it is true that the exchange of energy and signals across the interface is very much affected by the characteristics on both sides, one need not lose track of which side has which characteristics. However, which is the causing side and which is the affected side is arbitrary. (Does a force on a spring, for example, cause a displacement, or does a displacement cause a force?) The point is that this reciprocity/ambiguity lives quite happily in a Cartesian world of Newtonian physics. One need not throw up one's hands in despair concerning analysis about who is doing what to whom, but only admit that there are limits to breaking into the closed loop to identify causation.

A person on the street who sees a doorway into a building immediately perceives it as a means to walk into the building. Is that affordance in the environment, or is it a mapping from street to building embodied in the synapses of the person's brain—a model that has been built up by what the person has experienced? Both! And the dimensionality of doors and building in relation to the street is in the environment, and so too is it in the brain. But the physics of seeing and the physics of walking are still quite separable, even though they can be mapped one on the other.

The coupling of human actor and physical environment is dealt with in various ways by psychologists. First is the fact that all understanding of the physical world is mediated by human perception. The psychophysicist Stevens (1936) has argued that psychology is the most basic science. He uses the term *propaedeutic*, meaning "coming first", as his descriptor, psychology necessarily preceding physical science in the sense that all physical instruments are necessarily devised by people and their measurements are interpreted by people. Second is the now well-accepted experimental fact that economic rewards and benefits in the environment determine human perception. Green and Swets (1966), whose use of signal detection theory as a basis for understanding threshold phenomena, demonstrate that absolute (minimal-stimulus energy thresholds) are very much determined by the payoff for judging that a signal is there when it is really there, as compared to the penalty for judging it is there when in truth it is not. Finally, there is the demonstrated reciprocity of the human to environment in the stimulus-response dynamics of control. McRuer and Jex (1967) and colleagues have shown that human controllers quite naturally adopt a transfer characteristic that is the inverse of the dynamics of the system they are controlling.

All three of these examples of reciprocity point to a kind of ambiguity of causality as between the human and environment. But, here again, the reciprocity of actor and environment and the ambiguity of cause and effect are perfectly acceptable within a traditional Cartesian universe. Whether two physical systems are interacting—or whether one is animate and the other is not—makes no difference. There is a feedback loop. Analytical tools for understanding and modeling signals within feedback loops are well established. The whole process of estimation occurs within a feedback loop, but that does not negate the separate identity of model and environment being modeled.

The visual flow field is an interesting example of a phenomenon that is dependent on the closed loop, the perception–action dynamic. A Gibsonian might claim that the flow field is not in the eye or in the head or in the environment, but is an interaction. I would claim that it is in the brain synapses, a refined model of what it is like to move through an environment of objects.

Let us return for a moment to that ultimate challenge of understanding perception of (estimation of) reality, namely that of divine presence. The estimation paradigm, it seems to the author, provides a framework for rational discussion of God. Whether the perception is that of reality or is metaphor obviously depends on what model one starts with, what actions one takes as a result of what one observes (the control law), and what assumptions one makes regarding the sensory and action filters. What is virtual and what is real, what is subjective and what is objective, are a matter of one's criterion for modeling and believing the model. Hardly a new idea!

We can conclude that, as related to "presence" and "reality", the essential differences between the ontological assertions of Descartes and those of Heidegger and Gibson appear exaggerated, especially when viewed from the perspective of engineering estimation theory. That simple and mechanistic perspective may be helpful in confronting the ultimate questions of presence and reality.

Now to the views on reality by some professional philosophers.

Donald D. Hoffman, "Fitness beats truth". Hoffman, a cognitive scientist at the University of California Irvine, offers some very challenging theories about reality in his book *The Case Against Reality: How Evolution Hid the Truth From Our Eyes.* Hoffman argues that natural selection is necessarily directed toward Darwinian "fitness payoffs". Our perception of reality is based on internal models that are directed toward actions that better our situation in the natural world. He goes further and claims that when we look away, physical things are not really there. Ideas about these things, such as rocks, cars, and other people, are like icons on a computer desktop that we use to manipulate the world to improve our chance of our survival and our progeny. His mnemonic throughout the book is FBT, "fitness beats truth". Of many examples, he investigates animal behavior with respect to color, and shows us how color perception evolved to support behavior, and has scant connection to physical "truth". His ideas remind us of the internal model-based control concept of modern control theory described above, where an internal model continually adjusts itself to better its usefulness relative to a real world that inherently provides the fitness criteria.

Hoffman asserts that ultimately consciousness is the primary reality, "Conscious realism" is described as a *non-physicalist monism*: consciousness is the primary reality and the physical world emerges from that. In other words, what exists in the objective world, independent of one's perceptions, is a world of conscious agents, not a world of unconscious particles and fields. These particles and fields are icons in the user interface of conscious agents, but are not themselves fundamental elements of the objective world, he claims.

Are we living in a computer simulation? Nick Bostrom, a philosophy professor at Oxford University, in a 2003 paper by the above title (bold letters), argues that at least one of the following propositions is true: (1) the human species is very likely to go extinct before reaching a "posthuman" stage; (2) any posthuman civilization is extremely unlikely to run a significant number of simulations of their evolutionary history (or variations thereof); (3) we are almost certainly living in a computer simulation. It follows that the belief that there is a significant chance that we will one day become posthumans who run ancestor simulations is false, unless we are currently living in a simulation.

The Simulation Hypothesis, a book by best-selling author, MIT computer scientist, and Silicon Valley video game designer Rizwan Virk (2019), explains one of the most daring and consequential theories of our time. Drawing from research and concepts from computer science, artificial intelligence, video games, and quantum physics and referencing both speculative fiction and ancient eastern spiritual texts, Virk shows how all of these traditions come together to point to the idea that we may be inside a simulated reality like *The Matrix* (film). A recent TV series called *Upload* also conveys how telepresence and the metaverse are in our future.

The Simulation Hypothesis is the idea that our physical reality, far from being a solid physical universe, is part of an increasingly sophisticated video game-like simulation, where we all have multiple lives, consisting of pixels with the simulation's own internal clock run by some giant artificial intelligence. Simulation theory, according to Virk, explains some of the biggest mysteries of quantum and relativistic physics, such as quantum indeterminacy, parallel universes, and the integral nature of the speed of light.

Virk shows how the history and evolution of our video games, including virtual reality, augmented reality, artificial intelligence, and quantum computing could lead us to the point of being able to develop all-encompassing virtual worlds like the Oasis in the film *Ready Player One*, or the simulated reality in the film *The Matrix*. While the idea sounds like science fiction, many scientists, engineers, and professors have given *The Simulation Hypothesis* serious consideration. But *The Simulation Hypothesis* is not just a modern idea. Philosophers and mystics of all traditions have long contended that we are living in some kind of "illusion", and that there are "other realities which we can access with our minds".

Virk quotes prominent public figures and authors to support the credibility of his ideas:

> There's a one in a billion chance we are not living in a simulation.—Elon Musk
>
> I find it hard to argue we are not in a simulation.—Neil deGrasse Tyson
>
> We are living in computer generated reality.—Philip K. Dick (2018)
>
> The Simulation Hypothesis presents a radical alternative to current models of reality.—Jacques Vallée (1993)
>
> Riz Virk combines the mind of a scientist with the heart of a mystic, using video games to explain the virtual reality that we live in.—Dannion Brinkley (Brinkley at al, 1994)
>
> The Simulation Hypothesis is one of the few that could convince me that I probably live in a simulated universe. Encompassing the history of religions, philosophy, pop culture, physics and computers, Virk draws connections which show this theory is not only feasible but probably correct. If this sounds mind blowing, it is!!—Diana Walsh Pasulka (2021)

Is consciousness in everything? "We judge reality through our consciousness" argues philosopher Christof Koch (2012). Over the last decade, Koch has worked closely with the psychiatrist and neuroscientist Giulio Tononi. Koch advocates for a modern variant of panpsychism, the ancient philosophical belief that some form of consciousness can be found in all things. Tononi's *Integrated Information Theory* (IIT) of consciousness differs from classical panpsychism in that it only ascribes consciousness to things with some degree of irreducible cause–effect power, which could include the Internet "Thus, its sheer number of components exceeds that of any one human brain. Whether or not the Internet today feels like something to itself is completely speculative. Still, it is certainly conceivable", but does not include "a bunch of disconnected neurons in a dish, a heap of sand, a galaxy of stars or a black hole". He and Tononi claim that IIT is able to solve the problem in conceiving how one mind can be composed of an aggregate of "smaller" minds, known as the *combination problem*.

David Chalmers thoughts on simulation and reality. Philosopher David Chalmers has written a recent book titled *Reality Plus* (2022), with many observations about VR. He offers a number of insights that I paraphrase here, particularly with respect to simulation and reality. As already mentioned, one key question that philosophers have struggled with is: Are we living in a simulated world? How do we know what is real? Can we prove we're not living in a computer simulation? Everything around us could be simulated. Maybe we think our consciousness could never be simulated, much as Descartes implied when he asserted "I think therefore I am".

"What evidence would prove that a simulation is wrong?" he asks:

> If it is inconsistent, that is, self-inconsistent, that would be a clue. If we're in a perfect simulation it's hard to see how we could ever gain evidence of that fact. It has been asserted that computers cannot simulate everything, for example some say they cannot

> simulate quantum processes. In principle we can get evidence for and against various imperfect simulation hypotheses, which presumably have empirical consequences that we can test, at least in principle.

Chalmers points out that simulations can represent reality that has happened, or that could have happened but didn't, or that never could have happened. He notes that a thought experiment is a type of simulation. Many movies and films such as *The Matrix* are based on simulations that seem to go beyond reality. Some tough-minded scientists and philosophers claim that a perfect simulation hypothesis is meaningless because it's untestable. Remember that Karl Popper insisted that the hallmark of a scientific hypothesis is that it is falsifiable. However, we've seen that the simulation hypothesis is not falsifiable, because any evidence against it could be simulated.

With respect to virtual reality, Chalmers claims, the simulation hypothesis suggests we're living in a fully immersive virtual world that we experience all around us. (See also Bostrom, 1990). If we're in a virtual world we're in a fully immersive VR. But what is reality? Virtual realism is the thesis that virtual reality is genuine reality, with emphasis especially on the view that virtual objects are real and not an illusion. Chalmers coins the term *virtual digitalism*, meaning that objects in virtual reality are simply digital objects, or, roughly speaking, are structures of binary information, bits. This is a version of digital virtual realism where digital objects are real.

He cites five ways of thinking about what is real: (1) reality as existence; (2) reality as causal power, meaning it can affect things or be affected by things; (3) reality as mind independence, meaning it doesn't go away; (4) reality has non-illusory-ness, which means things are as they seem; and finally (5) reality as genuineness, or authenticity, which means something is a really specific thing. Chalmers summarizes the above checklist with other words and applies it to simulated reality: Does it really exist? Does it have causal powers? Is it independent of our minds? Is it as it seems? Is it genuine? If we're in a perfect permanent simulation, then the objects we perceive are real according to all five of these criteria, and are real and not an illusion.

What is virtual reality? The word virtual comes from the Latin word *virtus* meaning strength and power. Also, the root of the word virtual comes from the same root as virtue. Specifically addressing VR technology, Chalmers turns to virtual reality headsets: do they create reality? Current VR headsets create audiovisual immersion but not full-body immersion in an interactive sense. Interactive means objects in the environment affect one another as well as the user, and the environment is affecting each.

So how real is virtual reality? We can get a grip by asking are the objects inside virtual reality real? Chalmers applies his idea of virtual digitalism mentioned above: A virtual object is a structure inside a simulation. Virtual objects have certain causal powers to affect other objects and also affect users and so on. Digital objects really have causal powers, so virtual objects are digital objects.

Chalmers credits the psychologist Mel Slater for introducing the term *presence* for the sense of "being there", where Slater breaks down *presence* into two illusions; (1) the *place illusion* of being in a place in spite of the sure knowledge that you were not there and (2) the *plausibility illusion* that what is apparently happening is really happening even though you know for sure it is not. Chalmers adds the *body ownership illusion* that a certain virtual body or avatar is "my body". He mentions the 1984 novel *Neuromancer* by William Gibson, often credited with provoking hard questions about virtual reality. If VR is an illusion, it could be like looking in a mirror, which is like experiencing objects in virtual space.

One last comment by Chalmers concerns augmented reality. He notes that: a notebook or a computer can be an extension of the mind, so whether something is stored in the brain or in the notebook, it's still in the mind. He claims that objects in augmented reality are extensions of the mind in the same sense.

REFERENCES

Barfield W, Rosenberg C and Kraft C: Relationship between scene complexity and perceptual performance for computer graphics simulations, *Displays and Technology Applications* 11: 179–185, 1990.

Bostrom N: Are we living in a computer simulation? *Philosophical Quarterly* 53(211): 243–255, 1990.

Bracken C, Pettey M and Wu M: Revisiting the use of secondary task reaction time measures *AI & Society* 29(4): 533–538, 2014.

Brinkley D, Perry P and Moody R: *Saved by the Light*, New York, Google Books, 1994.

Chalmers D: *Reality Plus*, New York, Norton, 2022.

Crick F: *The Astonishing Hypothesis: The Scientific Search for the Soul*, New York, Charles Scribner's Sons, 1994.

Dennett D: Beliefs about beliefs, *Behavioral and Brain Sciences* 1(4): 568–570, 1978.

Dick T: *Conversations with Philip K. Dick*, New York, Google Books, 2018.

Draper J, Kaber D and Usher J: Telepresence, *Human Factors* 40(3): 354–75, 1998.

Edelman, G: *Bright Air, Brilliant Fire: On the Matter of the Mind*, New York, Basic Books, 1992.

Flach J: Personal communication, 2/24/99.

Flach J and Holdren J: The reality of experience: Gibson's way, *Presence Teleoperators & Virtual Environments* 7(1): 90–95, 1998.

Gibson J: *The Ecological Approach to Visual Perception*, Boston, MA, Houghton Mifflin, 1979.

Gibson W: *Neuromancer*, New York, Ace Publishers, 1984.

Green D and Swets J: *Signal Detection Theory and Psychophysics*, Hoboken NJ, Wiley, 1966.

Heeter C: Being there, the subjective experience of presence, *Presence: Teleoperators and Virtual Environments* 1(2): 262–271, 1992.

Heidegger, M: *Being and Time* (J. Macquarrie & E. Robinson, Trans.). New York, Harper Collins (Original work published 1927).

Held R and Durlach N: Telepresence, *Presence* 1: 109–112, 1992.

Hendrix C and Barfield W: Presence in virtual environments as a function of visual and auditory cues, in *Proceedings of the Virtual Reality Annual International Symposium* 95, (pp. 74–82). Piscataway, NJ: IEEE Computer Society, 1995.

Hoffman D: *The Case Against Reality: How Evolution Hid the Truth from Our Eyes*, London, Allen Lane, 2019.

Kalman R: A new approach to linear filtering and prediction problems, *ASME Journal of Basic Engineering* 82: 35–45, 1960.

Kant E, *Critique of pure Reason*, 1781.

Karnopp D and Rosenberg R: *Analysis and Simulation of Multiport Systems. The Bond Graph Approach to Physical System Dynamics*, Cambridge, MA, MIT Press, 1968.
Koch K: *Consciousness: Confessions of a Romantic Reductionist*, Cambridge, MA, MIT Press, 2012.
McRuer D and Jex H: A review of quasi-linear pilot models, in *IEEE Transactions on Human Factors in Electronics*, HFE8(3): 231–249, September 1967.
Montavoni G and Riva R: Real presence: How different ontologies generate different criteria for presence, telepresence, and virtual presence, *Presence Teleoperators & Virtual Environments* 8(5): 540–550, 1999.
Mühlbach L, Boucker M and Prussog A: Telepresence in video communications: A study on stereoscopy and individual eye contact, *Human Factors* 37: 290–305, 1995.
Pasulka D: *American Cosmic: UFOs, Religion, Technology*, London, Oxford, 2021.
Schloerb D: A quantitative measure of telepresence, *Presence* 4: 64–80, 1995.
Sheridan T: Further musings on the psychophysics of presence, *Presence* 5: 241–246, 1996.
Sheridan T: Musings on telepresence and virtual presence, *Presence: Teleoperators and Virtual Environments* 1(1): 120–126, 1992a.
Sheridan T: Space teleoperation through time delay: Review and prognosis, *IEEE Transactions on Robotics and Automation* 9(5): 592–606, 1993.
Sheridan T: *Telerobotics, Automation, and Human Supervisory Control*, Cambridge, MA, MIT Press, 1992b.
Sheridan T: *What is God, Washington DC, New Academia*, 2014.
Slater M and Usoh M: Representation systems, perceptual position, and presence in immersive virtual environments, *Presence* 2: 221–233, 1993.
Slater M, Usoh M and Steed A: Depth of presence in virtual environments, *Presence*, 3(2): 130, 1994.
Steuer J: Defining virtual reality, dimensions determining telepresence, *Journal of Communications* 42: 73–93, 1992.
Stevens S: Psychology: The propaedeutic science, *Philosophy of Science* 3: 90–103, 1936.
Tononi G: Integrated information theory of consciousness: An updated account, *Archives Italiennes de Biologie* 150(4): 293–329, December 2012.
Valee J: *Forbidden Science: Journals 1957–1969*, Berkeley, CA, North Atlantic Books, 1993.
Vick R: *The Simulation Hypothesis, An MIT Computer Scientist Shows Why AI, Quantum Physics and Eastern Mystics All Agree We Are In a Video Game*, New York, Google Books, 2019.
Witmer B and Singer M: *Measuring Immersion in Virtual Environments, Tech. Report 1014.* Alexandria, VA: U.S. Army Research Institute for the Behavioral and Social Sciences, 1994.
Witmer B and Singer M: Measuring presence in virtual environments: A presence questionnaire, *Presence: Teleoperators and Virtual Environments* 7: 225–240, 1998.
Zahorik P and Jenison R: Presence as being-in-the-world, *Presence: Teleoperators and Virtual Environments* 7: 78–89, 1998.
Zeltzer D: Autonomy, interaction, and presence, *Presence: Teleoperators and Virtual Environments* 1(1): 127–132, 1992.
Zsambok C and Klein G (Eds.): *Naturalistic Decision Making*, Hillsdale, NJ, Lawrence Erlbaum, 1997.

4 Applications

Telepresence applications are discussed in this chapter under 16 headings. Some applications primarily involve robotic telepresence, some primarily involve virtual reality, and some involve both, as will be explained. In each application, some sense of telepresence by the human operator is critical. When local autonomy is used in remote robot function, one can consider telepresence much the way a parent wants to be present for a child's piano recital or sports performance. There is also the need to monitor the automatic execution of programmed tasks under human supervisory control, as described in Chapter 2.

At the end of this chapter, I will mention an idea called the "metaverse". It is a vision for a future network of 3D virtual worlds focused on social connection. In futurism and science fiction, it is often described as a hypothetical iteration of the Internet as a single, universal virtual world that is facilitated by the use of virtual and augmented reality headsets, software, and related technology (*Time Magazine*, 2021).

EXPLORATION OF SPACE

Space probes can be said to be man's most exotic form of telepresence. In the 1960s, human astronauts were physically sent to the moon because that was the only way to "be there". Robotics was in its infancy and virtual reality did not exist. Now, however, space exploration has changed dramatically, and robotic telepresence vehicles have replaced human astronauts. The public is well aware of NASA's space probes and lunar roving vehicles as well as those of other nations—on the Moon and Mars and elsewhere in outer space. Fong et al. (2013) review the control issues encountered. Because of the sophistication of NASA's contribution, I mention NASA-provided descriptions of seven of these:

1. **Curiosity** (Figure 4.1) is a mobile laboratory that was launched from Cape Canaveral in 2011. It landed on the Mars surface on August 6, 2012. This was the largest rover NASA has put on Mars, being twice as long and five times as heavy as its processors. Despite the extra size, Curiosity took many design elements from the previous generation of Mars rovers, such as a six-wheel drive, rocker–bogie suspension, and cameras mounted to the mast of the rover to help the mission's team direct the rover. However, unlike the previous generation, Curiosity contains an entire inboard laboratory for analyzing the soil and rocks on Mars. NASA engineered Curiosity to be capable of rolling over obstacles up to 65 cm high and traversing up to about 200 m per day on Martian terrain. Curiosity gets its electrical power from a Radioisotope thermoelectric generator. Other Mars landers Sojourner, Spirit, and Opportunity are no longer operational.

DOI: 10.1201/9781003297758-5

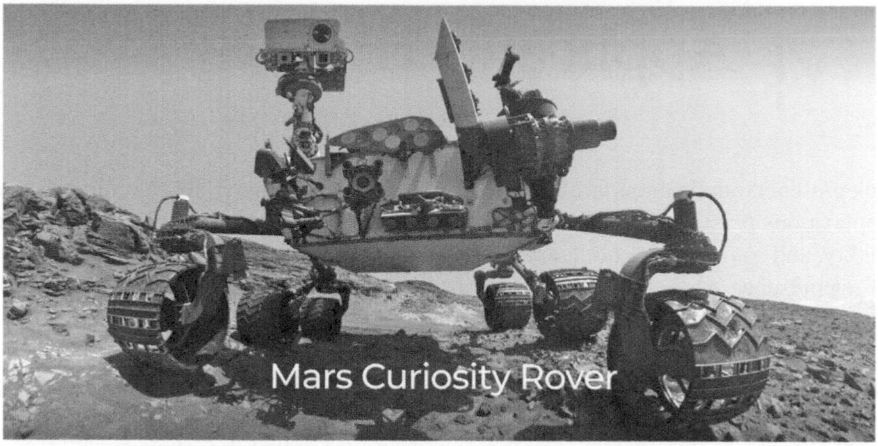

FIGURE 4.1 Mars Curiosity Rover (NASA).

2. **Robonaut** is a joint DARPA–NASA project designed to create a humanoid robot, which can function as an equivalent to humans. The large goal of the Robonaut project is to build a robot with a dexterity that exceeds that of a suited astronaut. Currently, there are four different robonauts with others in development, and this variety of robonauts allows for the study of different stages of mobility and tasking for each situation. All four versions of this robot use various locomotion methods. Robonaut uses telepresence and various levels of robotic autonomy. While not all human range of motion and sensitivity has been duplicated, the robot's hand has 14 degrees of freedom and uses touch sensors at the tips of its fingers. One of the benefits of a humanoid robot is that it would not have to need a whole new set of tools.
3. **Rassor** stands for Regolith Advanced Surface Systems Operations Robot. It is a lunar robot that will autonomously excavate soil when it is near completion, with its small tank-like chassis with a drum excavator on either side, mounted on arms which can help the robot climb over obstacles that may be in its way. With these arms, the robot can successfully right itself if it flips over and lifts itself off the ground to clear its tracks of debris. With the drums positioned vertically, RASSOR stands at about 2.5 ft tall and is expected to weigh about 100 lb. With an average speed of about 20 cm/s (five times faster than the Curiosity rover's top speed on Mars), the Rassor will work 16 hr a day for many years (a minimum of 5 years as stated in the design requirements). In its design, NASA has moved away from its usual fragile and slow robot to design something more robust and hardy. The two excavating drums are designed to slowly remove soil into a hopper that can hold 40 lb of material. The little robot will then drive to a processing plant where the lunar soil could be chemically broken down and converted into rocket fuel, water, or breathing air for astronauts working on the moon and even possibly Mars. In situ resource utilization of lunar soil for fuel could

save the costs of launching a rocket, as 90% of the rocket's weight consists of propellants.

4. **Spidernaut** is an arachnid-inspired Extra Vehicular Robot (EVR) that is being designed by NASA for construction, maintenance, and repair projects in future space missions that would be too difficult or too dangerous for a human. The Spidernaut's legs can move at three different points: one rotary joint in the hip and two more joints that are linearly actuated. Each leg weighs 40 lb but is capable of supporting 100 lb and exerting upwards of 250 lb of force. With the robot's final weight of nearly 600 lb evenly spread out across its eight legs, Spidernaut will be able to climb across many surfaces including solar panels and the exterior of spacecraft without causing any damage. The feet of the robot are modular, meaning they can be removed and replaced for different situations that the robot may be placed in. The avionics and other electrical systems of Spidernaut are located in what would be its thorax and are made up of brushless DC motor controllers and power and data distributors along with the power source. The robot is powered by a 72 V/3.600 W-hr lithium-ion battery, which feeds a Power Conditioning and Distribution Module that down-converts the 72 V main bus to all the needed voltages for all the different devices on board. NASA has also begun experimenting with a "web"-like cable deployment system that would allow the robot to climb and hang above structures that cannot support even light forces.
5. **ATHLETE** (All-Terrain Hex-Legged Extra-Terrestrial Explorer) is a six-limbed robotic lunar rover test-bed that is being developed in the Jet Propulsion Laboratory (JPL) at the California Institute of Technology. ATHLETE is a test-bed for various systems that could be used for lunar or Martian exploration. Each of the ATHLETE's six limbs has six independently operated joints. For general traveling purposes, the ATHLETE rolls on its six wheels, but if it encounters more rugged and extreme terrain it is able to lock each wheel into place and walk using its limbs. The first-generation ATHLETE was developed in 2005 and consisted of six 6-degree-of-freedom limbs mounted to the frame of the robot. With a standing height of 2 m (6.5 ft) and a hexagonal frame of 2.75 m (9 ft), ATHLETE weighs about 850 kg (1,875 lb) and can carry a load of up to 300 kg (660 lb). Only two were ever constructed in 2005. One is still operational today.
6. **Spheres** stands for Synchronized Position Hold, Engage, Reorient, Experimental Satellites. These satellites are about the size of a bowling ball and have undergone extensive experiments by NASA. Each Sphere is self-contained with its own power, propulsion, computers, and navigation equipment. The Spheres are designed to be used inside a space station to test how well the spheres can follow a set of detailed flight instructions. While inside a space station, three spheres will be given a set of instructions such as an autonomous rendezvous and docking maneuver. The results from the sphere testing will be applied to satellite servicing, vehicle assembly, and future spacecraft that will be designed to fly in a formation.

7. **Pioneer** is a robot developed in response to the Chernobyl disaster to clear rubble, make maps, and acquire samples inside the Chernobyl Unit 4 reactor building. The Pioneer project is a collaboration with Carnegie Mellon University and other groups inside and outside NASA. The concrete sampling drill on the Pioneer is designed to estimate the material strength of the floors and walls while it cuts out samples for later structural analysis. Ownership was transferred to Ukraine in 1999 and Ukrainian experts began learning to use it.

Control of Mars spacecraft for Mars orbit. The teams in charge of the rovers currently on Mars (Opportunity and Curiosity) have to deal with 5–40 min delays in getting data back and forth because of the time it takes for signals to travel at the speed of light. The process can take even longer if the orbiters that help relay the data aren't aligned just right. When sending a command to the Curiosity rover, for example, it can sometimes take a day or longer for the robot to get the message, complete the task, and send the data back to Earth.

Lester et al. (2017) (Figure 4.2) suggest a safer and more cost-effective compromise than NASA's previous way of operating. The system would involve putting humans on an orbiting station and operating robots on the surface, at a distance close enough that the latency amounts to only a fraction of a second. The claim is that this is in many ways superior to an astronaut on the surface in a bulky pressurized spacesuit with limited consumables.

FIGURE 4.2 This shows how astronauts on an orbiting space station could control robots on Mars while staying in touch with Earth. The space station is not shown to scale (NASA).

The James Webb is the most ambitious NASA project to date. By viewing the universe at infrared wavelengths, Webb will show us things never before seen by any other telescope. It is only at infrared wavelengths that we can see the first stars and galaxies forming after the Big Bang. This is because these far-away objects are traveling away from Earth so fast that, by the Doppler effect, their wavelengths are long (hence infrared). Its technical characteristics are as follows:

Mission Lifetime	5 years (10-year goal)
Orbit	L2 (the Second Sun–Earth Lagrange Point), 1,500,000 km from Earth
Sunshield Dimensions	Approximately 22 m × 12 m
Primary Mirror	6.5 m diameter aperture
Wavelength Coverage	0.6 to >27 μm
Sky Coverage	100%
Operating Temperature	~45 K (–380°F)
Payload Mass	Approximately 6,500 kg
Science	First light, assembly of galaxies, the birth of stars, planetary systems, and the origin of life

UNDERSEA RESEARCH

For decades, video cameras have been lowered into the ocean, but that was the extent of telepresence in bodies of water. Undersea robotic vessels containing manipulator arms and cameras were first developed by the Navy in the 1960s. Today, there are many private firms developing them, as well as Oceanographic institutions such as Woods Hole in Cape Cod, MA, and Scripps in San Diego, CA (Figure 4.3).

Early undersea robots, such as the one that explored the newly discovered Titanic, were tethered to the mother ship on the surface (Figure 1.4). Both electrical power and communication signals (video and control of thrusters and manipulator) went through the tether. Today's undersea robots are mostly untethered (called *autonomous underwater vehicles* or AUVs), and as a result can swim to great depths unencumbered by a tether. They have explored under the polar ice cap and have reached the deepest point in the ocean at 11,000 m. Most are designed for up to 6000 m.

It was an AUV that recently found Shackleton's famous ship Endurance at 10,000 ft depth in the Antarctic's Weddell Sea, one of the iciest and most formidable places on Earth.

AUVs (such as those pictured in Figure 4.4) can glide from the sea surface to ocean depths and back. Others can stop, hover, and move like blimps or helicopters do through the air. Solar-powered AUVs can spend a portion of their time at the surface, blurring the distinction between undersea and surface vehicles. AUVs can run for as short as a few hours or as long as days or even months—depending on the power source and the task at hand—before the battery needs recharging.

With improvements in acoustic communication through water, it is now possible to transmit small amounts of data between AUVs and their operators. Vehicle status, such as depth, battery level, and even some sensor data, can be sent from the AUV

FIGURE 4.3 The James Webb ultraviolet telescope with the sun shield folded (before deployment) (NASA).

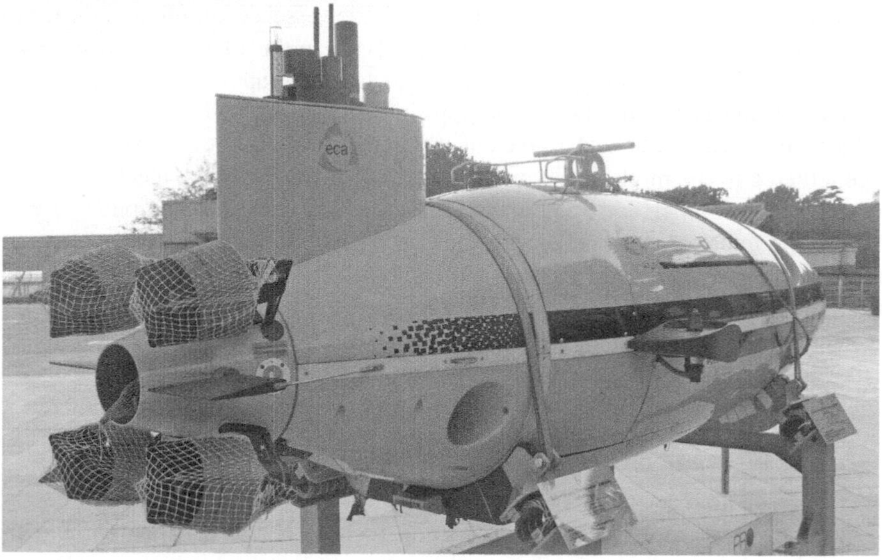

FIGURE 4.4 An autonomous undersea robot (Wikimedia Commons).

to the mission control computer onshore or on a boat. Low-level decisions, such as "speed up" or "turn", are made by the computer and software on the AUV, based on internal guidance sensors, but operators can transmit higher-level decisions, like "stop and photograph" and "come home". Operators can also change the survey area while an AUV is submerged. Even though AUVs may receive messages from an operator, the operator is not steering. This is an important distinction from tethered vehicles that are remotely operated continuously, which an operator can control by using a joystick.

One of the most promising areas of scientists' research using tethered vehicles as well as AUVs is the exploration of hydrothermal vents, crevices in the ocean floor. They first discovered hydrothermal vents in 1977 (https://www.whoi.edu/feature/history-hydrothermal-vents/index.html, 1977) while exploring an oceanic ridge near the Galapagos Islands. To their amazement, the scientists found that the hydrothermal vents were surrounded by large numbers of organisms that had never been seen before. The discovery of hydrothermal vents showed that life could thrive independent of the Sun.

Until recently, it could be said that the great majority of the Earth's biosphere was unexplored. This statement refers to the great expanse of ocean that lies between the few feet near the surface that have been well known to fish and aquatic plant science and the ocean bottom, more recently explored by manned and unmanned robotic vessels. Many new species of both plants and animals have recently been discovered at the intermediate depths.

TRANSPORTING PEOPLE AND GOODS

A self-driving automobile or truck is inherently a human-supervised telerobot. It has technology that senses its environment, computer-based programmable memory and control logic, and effectors that propel it along the road or brake it to slow or stop. Navigation by GPS, which also depends on detailed digital maps of roads and addresses, has been essential to this automation. Experience with self-driving cars suggests that human telepresence in the form of supervisory monitoring and control are life-savers. Evidence suggests that in predictable environments (like long, open highways), some degree of autonomy makes driving safer than 100% manual control. Much of the human factors research on human monitoring of boring unchanging tasks shows that humans get drowsy and inattentive after 30–40 min. How much human supervision is necessary is bound to change as automation improves and becomes more acceptable.

The prospect of semi-autonomy is especially attractive for long-haul trucking, a notoriously unattractive job. Especially at night, automated trucks could allow the driver to sleep and perhaps be monitored over a secure (encrypted) communication channel by a remote human dispatcher. Such a person might monitor a small number of such vehicles.

So too is a modern commercial aircraft a telerobot. Cross-country flying is mostly automated, the autopilot (now called the *flight management system*) having been programmed to fly at a prescribed airspeed between a series of waypoints and at a prescribed altitude. The pilots are there to manage takeoff, landing, and accommodate changes as directed by air traffic controllers, who track the aircraft on radar. Currently occurring technology improvements to air traffic control ensure safe separation between aircraft, avoidance of serious weather threats, and mitigation of unsafe congestion around busy airports.

In the early days of flying, large passenger aircraft were required to have three pilots: a senior pilot, a copilot, and a navigator. Several decades ago, technical advances and resulting easing of mental workload allowed the navigator position to be eliminated. There is a current debate over whether the copilot could be eliminated, or at least operate from a remote site, possibly playing safety backup pilot as needed, and possibly serving in that role for multiple aircraft. After all, many small aircraft, including air taxis, even those carrying passengers, already operate with a single pilot. These kinds of changes are being eyed especially for air freight operations, and, if adopted, would precede such changes for large passenger aircraft.

Railroad trains are the oldest and least automated vehicles for carrying passengers and freight. Some of the same additions of automation and "remoting" of human control could be considered for trains. Trains already have remote dispatchers, but they are not self-driving, except for automated rail systems that carry passengers between airport terminals in many cities. It is remarkable that even though trains are confined to tracks, they still experience overspeed derailments and crashes with other trains. In the future, additional sensors to monitor track conditions, stalled vehicles at crossings, and weather will help.

So-called rotorcrafts include both what are commonly called helicopters and vehicles with multiple rotors that are commonly called drones. The latter vehicles are now being produced in great quantities in small sizes and at low cost, mostly for amateur hobbyists. This boom in remotely controlled drones has been a headache for air traffic regulators (in the US Federal Aviation Administration, or FAA), which are making every effort to keep them from interfering with regular operations around airports and near crowds of people.

However, small teleoperated drones have a great future in performing small package delivery, and merchandisers such as Amazon are actively experimenting with how to do this. No doubt much more elaborate electronic maps and databases of addresses will be developed to enable this to happen, and there will be issues about hovering and landing drones in homeowners' private spaces (I will take up that issue in Chapter 5).

Ships too are being remotely controlled using telepresence.

The examples listed above involve transportation using discrete vehicles. But oil and gas are different, as they are transported in pipelines. One critical problem with pipelines is that they must be inspected for breaks, clogs, or other intentional damage. One way to do this is with a sensor-equipped robotic device that "swims" through the pipeline and detects problems and their location.

Lastly, we consider the application of telerobots in warehouses—to find and deliver items to locations to be assembled or packed into containers. Conventional warehouse telerobots are programmed to move to a location in front of the tall bin or set of shelves, where a robotic arm rises to the proper level, and a robotic hand reaches in to grasp the programmed item. Amazon has an interestingly different approach. Their robot is only a few inches tall. It is programmed to move underneath a shelf unit, pick up the whole storage unit, and bring it to the human worker who is doing the packaging, then return it to a holding area.

FLIGHT AND DRIVER TRAINING SIMULATION

Long before there was even terminology for virtual reality, flight simulators were being used to train pilots. Initially, these were just mockups of the cockpit, with the trainee running through takeoff and landing procedures while touching the proper button or control on the mockup or diagram. The author was involved in training the first crew of Apollo astronauts using such paper mockups of the control panel of the Lunar Excursion Module (LEM).

Gradually pilot training simulators became more sophisticated, to the point where the out-the-window visual scene (runway, inflight terrain) completely corresponds to what the pilot would actually see in a particular stage of flight. The pilot trainee operates actual controls and observes actual cockpit displays while experiencing an accurate dynamic sense of how the aircraft would respond. This is all contained in a cabin that itself moves in 6 degrees of freedom (on what is called a Stewart platform) to provide the pilot trainee the corresponding motion sense of roll, pitch, and yaw. Such a simulator is pictured in Figure 4.5. This can be called mixed reality since the visual scene is generated by computer graphics, while the technology to produce the proper motion experience is physically robotic. These flight simulators have become

FIGURE 4.5 A typical flight training simulator. The view out the windscreen is virtual (NASA).

so realistic that in some cases the pilot can go right from the training simulator to flying the actual aircraft with a full complement of passengers.

A similar type of mixed reality simulation is being used for research and training automobile and truck drivers. Figure 4.6 shows the National Advanced Driving Simulator (NADS) located in Iowa City, Iowa (https://www.nads-sc.uiowa.edu/drivingstudies.com/). It is a cabin similar to a flight simulator, with a computer-generated roadway scene corresponding to how the driver subject steers, accelerates, and brakes. Because lateral and longitudinal accelerations are normally greater in emergency highway vehicle handling than in flight, the whole of the cabin and Stewart platform are mounted on a very large X–Y movement base.

A high-resolution (5 million pixels) 360-degree visual scene is projected on the screen (15,000 polygons refreshed at 60 Hz) under program control. A correlated high-fidelity surround-sound audio system is also provided, as well as a control feel at 60 Hz.

Sophisticated software and computer architecture allow realistic accident scenes to be generated and coordinated with the vehicle motion, sound, and control feel.

Working with industry and government partners, the NADS research program has tested a range of driver warning systems for forward collision, lane departure, road departure, blind spot vision impairment, anti-lock braking failure, electronic stability control, and adaptive cruise control, as well as road signage and other compromised vision issues. This simulator has been used to investigate a variety

FIGURE 4.6 The National Advanced Driving Simulator (Wikimedia Commons).

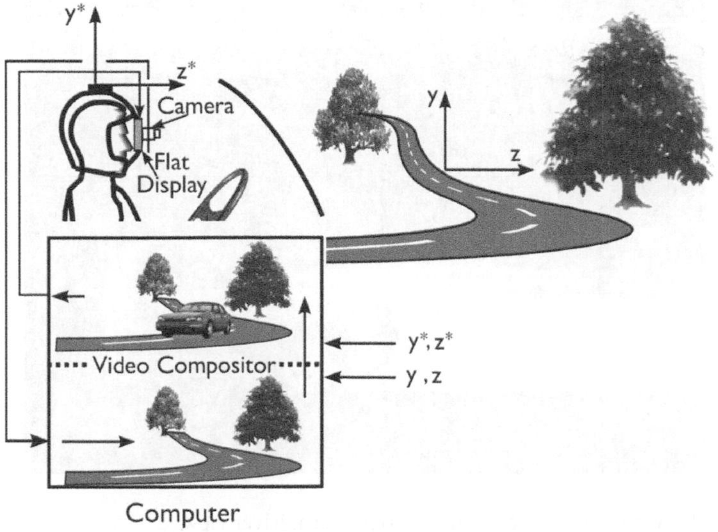

FIGURE 4.7 Augmented reality simulation of a near accident.

of questions about how aging, physical handicaps, highway conditions, other driver behavior, use of drugs, use of car phones, navigation systems, or other auxiliary devices while driving affect safety. It also assists in the engineering of new safety aids such as intelligent cruise control, radar collision-warning alarms, etc.

A driver of a simulation knows it is a simulation, so may not behave as she would in actual highway driving. This limitation in driver training can be overcome to some degree by a technique developed in the author's MIT lab. It exposes fledgling drivers in real vehicles to perceived near-crash situations without actual danger. This is done by having them drive an actual vehicle on a lonely road devoid of traffic and wear an augmented reality headset display. For an initial segment of the actual drive, everything is programmed to be normal, including the vehicle handling, vibration and engine noise, and all of the body sensations. But at some point, another vehicle starts to emerge from a side road, visually perceived as an impending collision, so the driver makes an emergency swerve or braking action. But there is no real collision. That vehicle is rendered by a computer in the augmented reality (AR) display.

Figure 4.7 shows the configuration of a test, and Figure 4.8 shows what the test driver actually saw. This test was conducted well before good AR displays were available, so the driver was viewing through a video display. Everything in the image is the observed real road except for the truck. The small white spot on the left side of the road was used to align the virtual image with the real road (Sheridan, 2007).

HAZARDOUS ENVIRONMENTS (FIRE, POLICE, RESCUE)

A very promising application of telepresence robots is for hazardous environments associated with fire rescue and related police work where building entry is dangerous.

FIGURE 4.8 Virtual vehicle emerging into forward driver view.

FIGURE 4.9 Los Angeles fire fighting telerobot (NASA).

Figure 4.9 pictures one such telerobot that has the capability of bringing a fire hose into a burning building, forcibly pushing furniture or other obstacles out of the way, and hopefully seeing or hearing a trapped victim in the fire.

Police operations sometimes require some means of entering a building where a gunman is threatening, and subduing (taser, teargas, actual shooting), or at least engaging such a person in conversation without endangering any lives. Such a device might take the form of a small tractor, or it could be a snake-like configuration, of a legged machine.

For getting live video in dangerous environments, a small drone-like device that carries cameras might be especially useful. It could fly into or through spaces that are dangerous for one reason or another. Small drones can carry sensors and software to be guided by a remote human operator, while at the same time automatically avoiding walls, ceilings, and floors, and simultaneously looking in multiple directions to allow human operators to see what the telerobot sees.

CLOSE INSPECTION AND MANIPULATION IN DANGEROUS ENVIRONMENTS

There are various telepresence tasks that involve close-up visual inspection coupled with specialized manipulations, tool use, or materials processing that might otherwise be done by a human directly but which are dangerous. These might include welding or inspection of buildings or bridges or other structures where getting positioned is awkward or dangerous for a human because of height and danger of falling; or handling dangerous chemicals or biological agents; or performing routine maintenance on machinery that is operating or at high temperatures or poses radiation or a collapse hazard such as in a mine.

If a telerobot is applied in such situations, it must overcome difficulty in accessing the work location, getting a stable base, applying proper forces, and making accurate robotic arm movements under close visual and/or tactile observation. More than six degrees of freedom for the robot hand or tool, and manipulating a camera on a separate multi-degree-of-freedom arm would be likely requirements.

Figure 4.10 shows a remotely controlled railroad track welder on a robot arm.

FIGURE 4.10 Railroad track welder on telerobot arm (Wikimedia Commons).

AGRICULTURE AND AQUACULTURE

Robotic telepresence is occurring in many aspects of agriculture. Remotely controlled machines rolling across the ground or in the air are already planting, watering, and harvesting rows of crops in fields, much as is depicted in Figure 4.11. They can also be used to observe livestock and confront animal predators or persons anticipating theft.

In urban areas, truck farming is occurring inside buildings with telerobotized machinery.

In lakes and ponds, feeding and harvesting fish is becoming robotized. In all such operations, it is essential to monitor the operations, which are done remotely using built-in sensors as well as means for periodic remote visual inspection by humans.

WIDE AREA SURVEILLANCE

Wide area surveillance on land has traditionally been done by human observers from aircraft. Increasingly this is being done by teleoperated drones. Figure 4.12 shows a photo of grasslands being surveilled with sensors discriminating different crops and grasses by contrasts (e.g., patterns and color) determined by the sensors. Such techniques can be used to estimate production volume and identify problems such as erosion, drought stress, and disease and provide verification for crop insurance. Rescue at sea is another application.

Drone surveillance is also replacing human building inspectors, who heretofore have had to make walking rounds of commercial and public buildings and grounds.

Remotely controlled drones are also being used for lidar photogrammetry (measurements of the timing of laser pulses reflected off the ground). A drone flying a trajectory at a fixed (barometric) altitude above sea level can map the topography.

Police use teleoperated drones in looking for prison escapees or detecting persons illegally crossing national borders.

FIGURE 4.11 Mechanized harvester feeds grain into truck as it moves along, both remotely controlled (Wikimedia Commons).

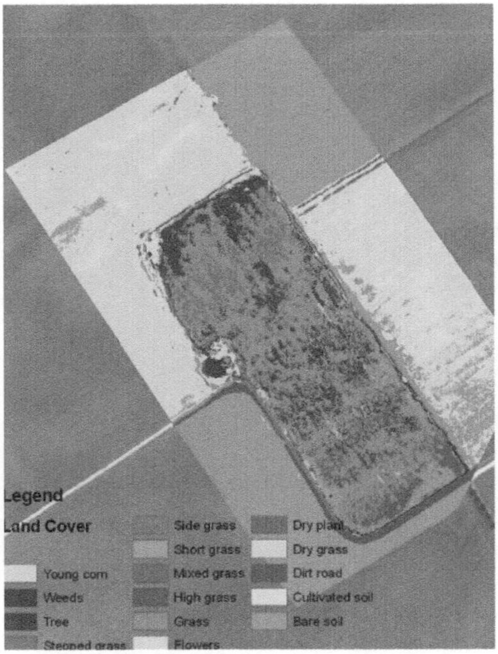

FIGURE 4.12 Image from a surveillance drone (US Dept of Agriculture).

TELEMEDICAL CONSULTATION AND TELESURGERY

Some medical consultation via telephone has always occurred. Sharing medical scan images via the Internet has also been common for a number of years, but it grew especially during the pandemic (Feizi et al., 2021). When Zoom and similar services came along it was immediately applied to medical consultation between patients and doctors (Figure 4.13). New developments in adapting cell phones for performing and communicating measurements of heart rate, blood pressure, blood oximetry, etc., are occurring, in addition to simply communicating photos of parts of the body.

Telesurgery began with the advent non-robotic minimally invasive surgical tools being used for gall bladder removals, often viewing the internal site through optical fiber bundles. Then came miniature video sensors, which provided a better picture. In 1995, the DaVinci robotic surgery system (https://www.davincisurgery.com/) was introduced, in which the surgeon views the wound site on a visual display and controls scalpels of other surgical instruments using hand controls such as shown in Figure 4.14. The robotic instruments and patient table are on the right between the nurses. One advantage of such telesurgery is that the scale of the surgeon's hand movements relative to the scale of movement of the surgical tool can be modified to suit the individual case, for example, extremely small and delicate movements can be executed by much larger hand movements of the surgeon. Further, the system can eliminate any jitter in the surgeon's hands (Atashzar and Patel, 2019).

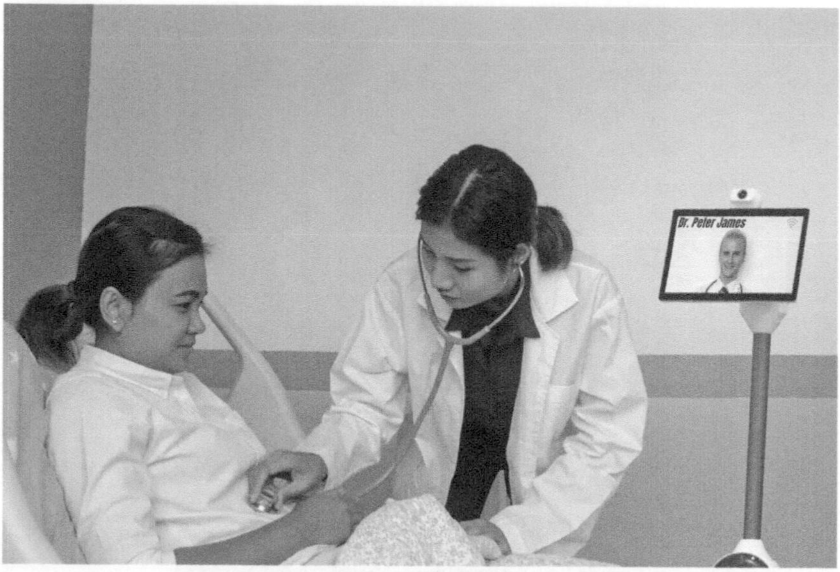

FIGURE 4.13 Physician in screen on right participating a nurse's examination of patient (NIH).

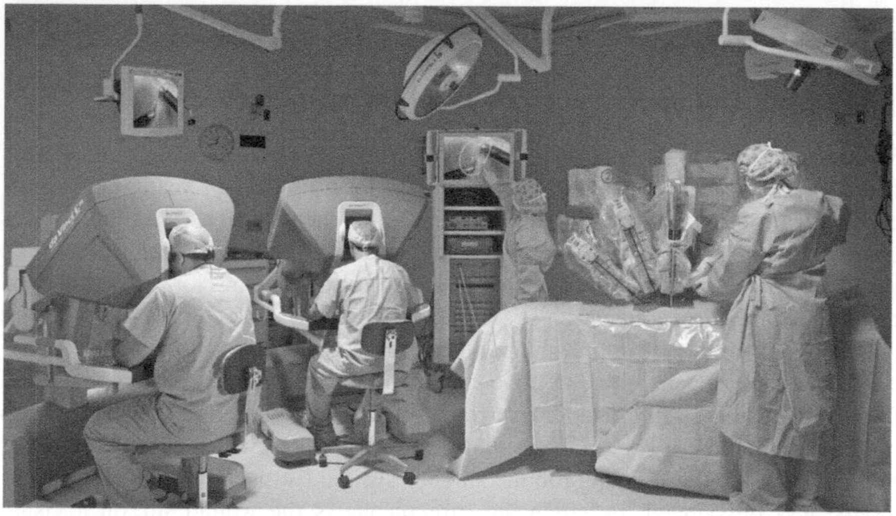

FIGURE 4.14 DaVinci surgical robot system (Wikimedia Commons). Two surgeons are participating in the procedure.

For many years, the US Army had faced the problem of soldiers bleeding to death on the battlefield before they could be brought to where a surgeon can stop the bleeding. For that reason, the Army experimented with the idea of telesurgery over long distances, where the robotic instruments in the field could be controlled by a surgeon at a rear position. Now that capability is being applied to accommodate emergency surgery on patients in remote rural locations far from where a surgeon is likely to be available.

A different approach (Figure 4.15) is a microsurgery system developed at NASA's Jet Propulsion Laboratory by Das et al. (1996, 1999). It is a miniature version of the old master–slave manipulator shown in Figure 1.1. JPL collaborated with MicroDexterity Systems Inc. to develop this robotic platform with important applications to medicine. This lightweight portable workstation was conceived to enable new procedures on the brain, eye, ear, nose, throat, face, and hand. The workstation controls include force feedback and other features to enhance manual positioning and tracking in the face of myoclonic jerks and tremors that limit most surgeons' fine-motion skills.

ASSISTANCE TO HANDICAPPED AND ELDERLY

Crude orthotic mechanisms have been in the medical appliance mix for many years, and simple artificial arms actuated by muscle (electromyographic, EMG) signals have been available. More recently, prosthetic arms actuated by EEG electrodes placed at key locations on the scalp or implanted in the brain have produced promising results for patients who otherwise would have no upper limb function.

So-called powered exoskeletons were first developed by the Army in the 1970s, the idea being to enable soldiers to carry much heavier loads than they normally could. At about the same time, engineers and medical specialists began using this technology to develop externally powered exoskeletons to help patients with lower extremity paralysis to stand and walk for short distances. Figure 4.16 shows an early

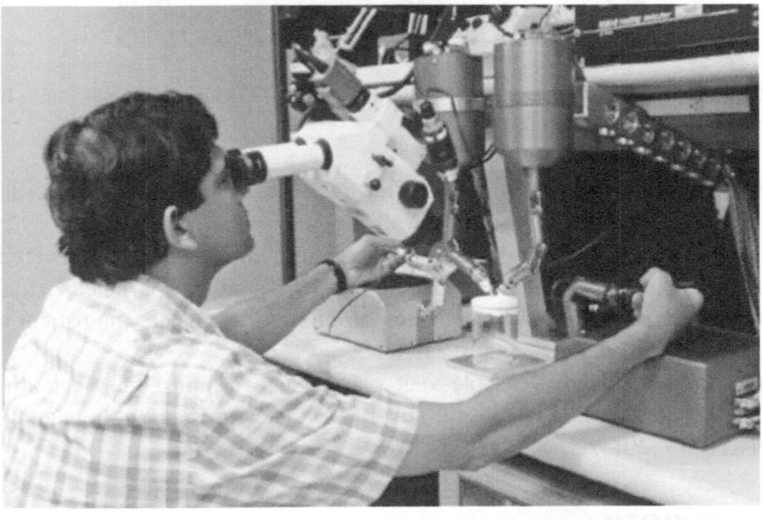

FIGURE 4.15 JPL microsurgery system. Each of the operator's hands is controlling a miniature force-reflecting 6-DOF arm, which in turn is controlling a miniature 6-DOF manipulator working on the white circular table in the center of the photo (NASA).

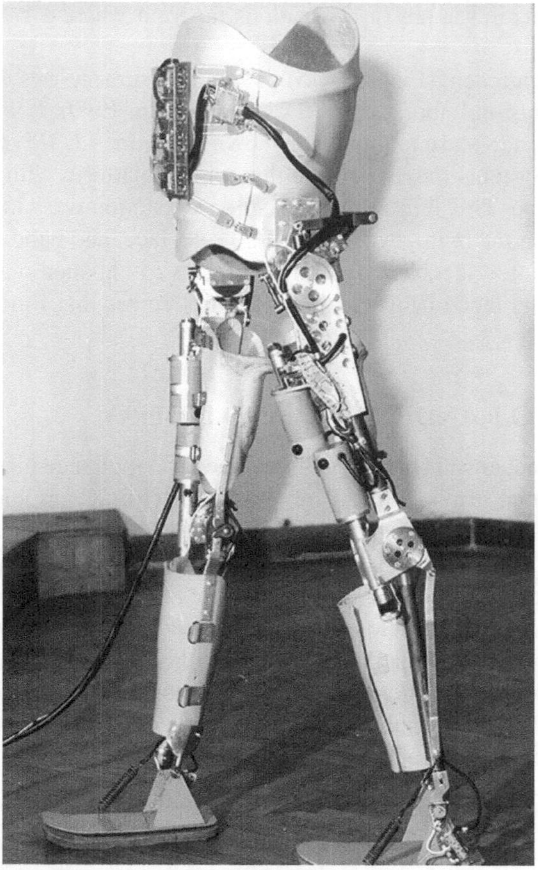

FIGURE 4.16 Externally powered exoskeleton made in 1974 at the Mihajlo Pupin Institute in Belgrade (Wikimedia Commons).

such exoskeleton from eastern Europe. These devices have evolved, but still have a long way to go to achieve comfortable walking.

An interesting use of virtual reality is to help individuals overcome severe fear of heights, open spaces, or animals, etc. By having such people experience virtual environments where the cause of the phobia is gradually introduced, the patient adapts, and the phobia can be alleviated over time.

Simply helping disabled people visit places such as museums that otherwise would be unavailable is a wonderful advantage of telepresence robots and virtual reality (Kelvery, 2014).

SOCIAL ROBOTICS

Social robots are devices with the appearance of pet animals and people that children and the elderly enjoy as companions (Figure 4.17). Their purpose is to serve a

FIGURE 4.17 Children with social robot (Wikimedia Commons). This robot in a train station educates children about train safety.

person in a caring or educational interaction rather than to perform a mechanical task. Typically, they embody electronics to talk or purr or otherwise respond to user voice or tactile responses based on Internet communication with a server located elsewhere. The Japanese were probably the first to develop a commercial market for social robots, but now they have been quite popular in many countries.

This author published a review of research on social robots (Sheridan, 2020). The literature emphasizes that facial expression, gesture, and touch are essential to the acceptance of both users and caregivers. Understanding the intentions and movements of users is critical to safe and effective interaction. Healthcare providers, who initially were skeptical of technology, have been very helpful in refining social robots.

CLOSE COLLABORATION WITH ANOTHER PERSON OR ROBOT

By close collaboration, I mean that two (or a small number of) people who are physically separated are not only speaking and hearing one another over a communication channel, but they are able to observe and manipulate objects at each other's locations or at a common location. They might be able to pan, tilt, and zoom cameras in remote locations. They might be able to point or draw on a common or separate surface. They might be able to operate a common robotic arm in order to demonstrate or teach/learn a manual skill. They might be viewing their own flat screens, or by wearing head-mounted displays be able to control pan, tilt, zoom, or control other variables at remote locations.

Functions such as those listed above might serve the purpose of teaching a manual skill, going over written text or drawings, or getting/giving repair service information, or actually carrying out some service. For the service operation, for example, a person at the location of the thing to be serviced (industrial machine, home appliance) could be wearing an AR headset and seeing the actual object, with the remote service person's avatar hand or tool pointing to certain locations on the object. Meanwhile, the service person might only be seeing a flat panel display(s) of the televised object with the pointer superposed.

GROUP MEETINGS

Almost everyone is familiar with meetings conducted over Zoom, Go-to-Meeting, or other purveyors of group meeting services Figure 4.18). There are protocols for raising hands to be called upon, voting, and superposing graphics (Powerpoint slides, spreadsheets, video clips) in presentations.

In the early 1970s, experiments were conducted in 200 group meetings (Sheridan (1975) including parent–teacher meetings, political forums, and community planning sessions using simple anonymous voting technology. It became clear that by using anonymous voting feedback technology as part of group meetings, people can explore their differing perspectives and experiences around issues that divide them and normally prevent friendly communication. The technique allows every participant to make an anonymous coded response to questions posed by the moderator or another participant, and to observe instantaneously a tally of how many people

FIGURE 4.18 Teleconferencing among four groups (Wikimedia Commons).

voted in each category. The technique permits a rapid appraisal of consensus and controversy; it allows participants to reveal their ignorance, deal with controversial questions without intimidation, and generally make the discussion more responsive to the real interests and needs of the group. In some cases, quantitative procedures were used to rate alternatives against criteria and determine group preferences. Mostly these procedures should be regarded as an augmentation of normal free discussion and idea formulation, rather than a means for commitment to final decision. Figure 4.19 shows an example of questions used in one meeting, recorded by hand

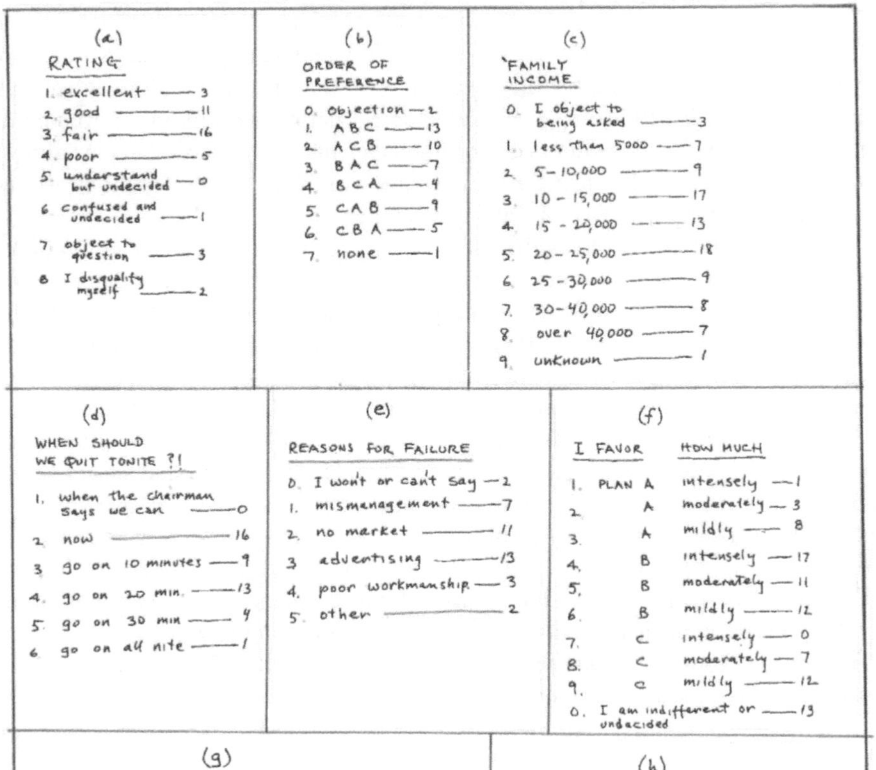

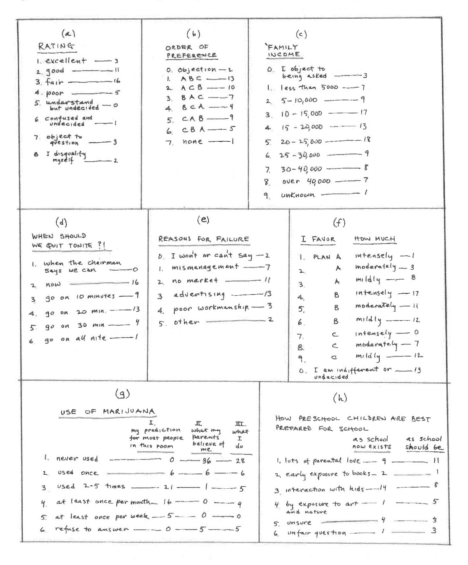

FIGURE 4.19 Examples of questions posed in group anonymous voting session in a parent–teacher meeting ass (Sheridan, 1975).

on an overhead projector as the meeting progressed and vote tallies came into the moderator. Keeping such a meeting very informal, with an opportunity to discuss the implications of each anonymous tally, and be able to modify the wording of the question and revote, etc., proved to be very important.

ARCHITECTURE AND DESIGN

One of the first dramatic demonstrations of virtual reality was to transform architectural drawings into graphic images that give the building client a sense of how the building will appear from various viewpoints, inside or out. Then this information was put together to form a 3D model that allowed a person to go inside. Now, wearing a head-mounted display, the observer can walk through the building at her own pace, look in any direction, and see what is there—before the building is even built. Color changes, relocation of doors or furniture, etc., in the model can be done almost instantly. As better touch-sensing gloves become available, observers will be able to reach out and "touch" the virtual walls and cabinets.

A related technique involves the use of laser photogrammetry, whereby a laser device carried on an aerial drone or on a vehicle on the earth's surface can map out the set of distances from any point to all other points within range of the laser. The result is called a *point cloud*. Multiple such maps can be stitched together to provide 3D models. This enables any existing structure to be turned into an accurate computer model, which is then useful for a variety of purposes, such as that above.

The trend these days is to keep an up-to-date *digital twin* for any building or other large civic project. A digital twin is a virtual representation that serves as the real-time digital counterpart of a physical object or process, and that also connects to a database. The first practical definition of digital twin originated from NASA in an attempt to improve the physical model simulation of spacecraft in 2010 (Grieves, 2019). Digital twins are the result of continual improvement in the creation of product design and engineering activities. Product drawings and engineering specifications progress from handmade drafting to computer-aided drafting/computer-aided design to model-based systems engineering and digital twins.

EDUCATION

Virtual classes have an important place in education. However, based on experience during the Covid pandemic, compulsory virtual classrooms do not have a good reputation with elementary school children and their parents. The children obviously missed the direct and free interactions with the teacher and each other. This forms a challenge for educators and human factor engineers.

For higher-level education, the international Internet online learning organization EdX tells a different story (https://impact.edx.org/2022). Based on digital courseware initially developed by MIT and Harvard, but now with 160 cooperating universities, over 3,000 courses are being offered globally in many languages in 196 countries. Roughly 400,000 students are enrolled at any one time. Over 6 million have been enrolled in credit-backed or credit-eligible courses. In a survey half of the students taking EdX courses claimed that EdX changed their lives, and 76% said EdX gave them access to an education that otherwise was unattainable. The top courses are computer science, data science, business, health care, and communications. Recently EdX has joined with 2U, an organization promoting online education at an even grander scale, with many more cooperating institutions.

There are obvious differences between the above-cited negative response to compulsory virtual classes for elementary school students and online courses for higher education. These include student age, motivation of the students due to maturity, and sophistication of the courseware. Online education has proven its value for motivated adults and higher-level courses (Millonig, 2014).

The use of computer graphics and virtual demonstrations are obvious components for teaching math, science, engineering, and economics, while group seminars using Zoom are probably best for humanities classes. In the future, head-mounted displays will surely be useful to embed the learner within virtual simulations, such as moving around within a molecule, cell, or within the human body. The learner might also be in an interactive situation with a person or group using hateful speech or experiencing an environment decimated by climate change. Being immersed and interacting within such virtual environments can be much more compelling than being a passive observer watching a film. The possibilities are exciting.

TOURISM

One of the nicest applications of virtual reality (VR) allows people to tour museums and historical and scenic sites anywhere in the world—without physically traveling at all. What is required is that someone produces that tour using a VR camera, which records what would be seen by looking in any direction as the photographer moves through the building or area. In the VR application, the (very large) set of stored information is stitched together in space and time, so that the eventual "tourist" can walk at any speed and turn her head in any direction to see what is there.

Because I am an enthusiast for this application, I take the liberty of mentioning a number of tours (or sets of tours) that can be played on a VR headset such as Oculus 2: Wander, Oculus TV, Brink Traveler, Rebuilding Notre Dame, Blueplanet VR Explore, National Geographic Explore VR, Alcove, Anne Frank House VR, and Ecosphere.

A future possibility is for the tourist to rent a mobile robotic vehicle that carries a camera adjustable in pan, tilt, and zoom, accompanied by an audio explanation if available, and drive herself through the museum. This allows for more freedom in movement compared to the recorded virtual tour described above (Starr, 2014).

SHOPPING AND MARKETING

Shopping by computer has been replacing going to physical places called shops. This is a trend that is continuing and is being enhanced by both VR and telerobotics. Computer VR techniques such as those used by Amazon allow the shopper to view the range of products from different angles and compare prices and technical specifications. This is clearly more convenient than traveling around to different shops, and it saves time and energy.

But there is a trade-off. Hardly any computer user can deny that their name, address, and shopping preferences are already in the database of Amazon, Google, and other high-tech firms with which they shop or do business. The panoply of techniques for attention-grabbing is only increasing. Furthermore, many smaller shops, both stand-alone and in malls, are going out of business.

But not quite. Although Door-Dash, PeaPod, and similar food delivery firms are growing, most people continue to want to go to supermarkets to see, touch, and collect food and related consumer goods. In such places, VR and robotic devices are helping people find items and take advantage of price computation and "sales". For example, robotic devices are roaming the aisles greeting customers, announcing sale items, and answering questions.

Marketing will continue to move beyond the home computer and the shopping mall. Marketing in airports, for example, has steadily grown in recent decades. Gas stations have become mini-malls. TV programming and newspapers will continue to harangue us with advertisements. Smartphones in our pockets remind us of products and services if we let them. The future is bound to see a battle between proponents of using technology to market and sell, and opponents who just want to be left alone as they shop.

GAMING AND THEATER

A video game or computer game is defined as an electronic system that involves interaction with a user interface or input device—such as a joystick, controller, keyboard, or motion-sensing device—to generate a visual experience with feedback. Now games are played on smartphones, computers, TV sets, and head-mounted displays like Oculus and HoloLens. The computer-gaming market has been an economic driver of virtual reality since its beginning. The first consumer video game was the arcade video game *Computer Space* in 1971. In 1972 came the iconic hit arcade game *Pong*, and many others soon followed.

The video gaming industry was estimated to be worth $178.73 billion in 2021, which was an increase of 14.4% from 2020. Recent forecasts are estimating the video gaming industry to be worth $268 billion by 2025. Asia leads the way in terms of the volume of gamers; it was found that Asia accounted for 1.48 billion gamers out of a total of 3.2 billion gamers worldwide. Put another way, 45% of gamers are based in Asia with gamers based in Europe taking second place at 22% (or 715 million) of the total (ew 2021). According to Wikipedia, one game called Fortnight had 350 million registered users in May 2020, Minecraft had 238 million, and Grand Theft Auto had 160,000.

Most computer games feature shooting, fighting, driving, or flying fast vehicles, and generally destroying people and things. Unfortunately, such virtual violence is where the current market is, and obviously that market is largely teenagers. There is extensive literature on the relation of computer game playing and aggressive behavior of teens, some authors claiming a strong relation and others claiming none.

The challenge is to utilize gaming for positive educational purposes, for example, using multiplayer games to promote community building, inter-group appreciation of cultural values, and improved mutual understanding by diverse groups.

MILITARY APPLICATIONS

Military institutions have traditionally been among the biggest funders of telerobotics and virtual reality research. They were mostly responsible for developing the Internet and have been behind many advances in robotics, computers, sensors, displays, and control.

Some military developments have genuine humanitarian functions. For example, unexploded land mines buried on former battlefields such as in Vietnam and Cambodia have killed a large number of people, especially children. Traditionally, such mines have been detected by dogs or by human volunteers tediously probing the ground at suspected locations at 45-degree angles (land mines are triggered from above). More recently ground-penetrating radar vehicles and other new robotic technologies helped to detect and dispose of these.

But the main job of the military is to intimidate adversaries and proceed to kill and destroy property to the extent that government authorities direct them to. There are various "smart" weapon systems that involve teleoperation by human operators and telepresence to monitor their use. For example, it is well known that pilots at Air Force bases in Nevada control large drones in war zones. The drones fly to and hover over what ground spotters think are appropriate targets, and smart missiles are aimed and released to blow up the targets. Such operations are deemed advantageous for the military because pilots are not endangered, but they are powerfully resented by people in nations where they are used, such warfare is deemed "unfair" because the pilots are not putting their own bodies at any risk. Much smaller drones, which carry both a camera and some explosives, were controlled by soldiers/operators on the Ukraine battlefield. These were directed to a target through a simple video display. Such robotized weaponry is surely the future of warfare.

THE "METAVERSE"

At the end of this chapter, after discussing the many applications of robotic telepresence and virtual/augmented reality, I feel compelled to comment on the planned ("hyped" might be a more appropriate term) "metaverse". The term "metaverse" describes a fully-realized digital world that exists beyond the one in which we live. The metaverse is envisioned as a representation of an immersive 3D virtual world where users can interact in different spaces by using their digital avatars. As in the real world, the metaverse can allow users to move around to different metaverse spaces. In fiction, a utopian metaverse may be portrayed as a new frontier where social norms and value systems can be written anew, freed from cultural and economic sclerosis. But more often metaverses are a bit dystopian—virtual refuges from a fallen world.

Metaverse was coined by Neal Stephenson in his 1992 novel *Snow Crash* (Stephanson, 1992) and the concept was further explored by Ernest Cline in 2011 in his novel *Ready Player One.*

Various metaverses have been developed for popular gaming use such as virtual world platforms like *Second Life* (Wikipedia). Some metaverse iterations involve integration between virtual and physical spaces and virtual economies, often including a significant interest in advancing virtual reality technology. Matthew Ball, a venture capitalist and prolific essayist, describes the metaverse not as a virtual world or a space, but as "a sort of successor state to the mobile internet"—a framework for an extremely connected life (https://www.matthewball.vc/all/forwardtothemetaverseprimer). "There will be no clean 'Before Metaverse' and 'After Metaverse'" he writes.

> "Instead, it will slowly emerge over time as different products, services and capabilities integrate and meld together. The metaverse is not an App Store with a catalog of titles. In the metaverse, you and your friends and your appearance and cosmetics can go from place to place and have different experiences while remaining connected to each other socially."

The term *metaverse* has seen considerable use as a buzzword for public relation purposes—to exaggerate development progress for various related technologies and projects. See, e.g. Clark, "What is the metaverse and why should I care?" *Time* December 29, 2021. Facebook (Meta) even changed its name in anticipation of its developments in this commercial application space. Information privacy, user addiction, and user safety are some of the concerns within metaverses, stemming from challenges facing the social media and video game industries as a whole.

REFERENCES

Atashzar S and Patel R: Teleoperation for minimally invasive robotics-assisted surgery, *Encyclopedia of Medical Robotics* 1: 341–372, 2019.

Ball M: The Metaverse. New York, Liverlight, 2022.

Clark P: What is the metaverse and why should I care? *Time December* 29, 2021. https://www.matthewball.vc/all/forwardtothemetaverseprimer.

Cline E: *Ready Player One*, New York, Random House, 2011.

Das H, Ohm T, Boswell C, et al.: Telerobotics for microsurgery, in *Proceedings of the 1996 IEEE Engineering in Medicine and Biology 18th Annual International Conference*, October–November, 1996, Amsterdam, The Netherlands.

Das H, Zak H, Johnson J, et al.: Evaluation of a telerobotic system to assist surgeons in microsurgery, *Computer Aided Surgery*, 4(1): 15–21, July 1999.

Feizi N, Tavokoli M, et al.: Robotics and AI for teleoperation, tele-assessment, and tele-training for surgery in the era of COVID-19; existing challenges, and future vision, *Frontiers Robotics AI*, 14 April 2021. https://doi.org/19.3389/frobt.2021.610677.

Fong T, Zumbado J, Currie N, et al.: Space telerobotics: Unique challenges to human–robot collaboration in space, *Reviews of Human Factors and Ergonomics*. Sage. December 23, 2013.

Grieves M: Virtually intelligent product systems: Digital and physical twins, in *Complex Systems Engineering: Theory and Practice*, S. Flumerfelt, et al., Editors. American Institute of Aeronautics and Astronautics, pp. 175–200, 2019.

https://impact.edx.org/2022. (See Education, p. 80).
https://www.davincisurgery.com/. (See Fig 4.14, p. 74).
https://www.matthewball.vc/all/forwardtothemetaverseprimer. (See Metaverse, p. 84).
https://www.nads-sc.uiowa.edu/drivingstudies.com/. (See Flight and Driver Training Simulation, p. 68).
https://www.whoi.edu/feature/history-hydrothermal-vents/index.html (1977) (see Undersea, p. 65).
Kelvey J: A quick reminder that technology can be wonderful; telepresence robots make it possible for people with disabilities to visit museums, *Slate*, July 22, 2014.
Lester D, Hodges K and Anderson R: Exploration telepresence: A strategy for optimizing scientific research at remote space destinations, *Science Robotics*, 2(7), June 21, 2017.
Millonig L: 'Telepresence robots' connect virtual teachers with classrooms, *Education Week* February 3, 2014.
Sheridan T: Community dialog technology, *Proceedings of the IEEE*, 63(3), March 1975.
Sheridan T: A review of recent research in social robotics, *Current Opinion in Psychology* 36, 7–12, 2020.
Sheridan T: *Vehicle Operations Simulator with Augmented Reality*. US Patent 7246050, July 2007.
Starr M: Explore Britain's Tate Museum after dark via robot, *CNet* August 12, 2014. Available at www.cnet.com/news/explore-britains-tate-museum-after-dark-via-robot.
Stephanson N: *Snow Crash*, New York, Bantam Books, 1992.

5 Challenges for Robotic Telepresence and Virtual Reality

A major motivation for me to write this book was concern about the future of telepresence, both of the real robotic variety and of virtual reality (VR) and augmented reality (AR). I cannot claim to be alone in these concerns; I find that they are shared by many people as the shape of these technologies emerge over the next decade or so. Below I discuss these concerns under 11 categories. Some examples are for robotic telerobots, for others the concern is associated with virtual/augmented reality and the anticipated "metaverse".

First I want to mention three experiences, call them mental images, that I acquired a great many years ago and that I feel apply now to the subject matter of this book.

With regard to robots and telepresence, I recall vividly some discussions that occurred during meetings (I was a participant) in conjunction with the mid-1960s' telerobotics project sponsored by NASA and the Atomic Energy Commission that resulted in three reports, E. Johnsen and W. Corliss: (1967) *Teleoperators and Human Augmentation*, NASA SP-5047; (1968) *Teleoperator Controls*, NASA SP-5070; and (1970) *Advancements in Teleoperator Systems*, NASA SP-5081. What was the future of telerobotics? One answer was that hostilities between nations would take the form of different adversaries implementing telerobots designed to battle for technological superiority in space, undersea, or on land, supposedly with no bodily harm to humans. We are not there yet, but clearly we are moving in that direction. Heretofore, soldiers in battle necessarily placed their bodies in jeopardy, but nowadays we are seeing telerobots wreaking destruction on others via telepresence, and it almost becomes easy and impersonal. Aerial drones can be piloted from arbitrarily far away so the pilots are not endangered, but can release laser-guided missiles that are killing people. I have always felt revulsion, and I sense that many people agree that such hostilities are immoral.

Nowadays some military contractors are pushing what they call *lethal autonomous weapons*, weapons that make decisions with essentially no human involvement. They are being tested at an accelerated pace, with the claim of "advantage", not only of keeping human soldiers out of harm's way but of making faster decisions than humans could make. But clearly autonomous robotic weapons lack compassion, empathy, and judgment in deciding when to "pull the trigger", with no human telepresence at all.

With regard to virtual reality, a second image that has stuck with me occurred when my wife and I were on sabbatical in Israel in the mid-1990s, when smartphones had just become available. We happened to be having dinner in a cozy restaurant,

DOI: 10.1201/9781003297758-6

seated next to a table with a young couple, who might well have been speaking lovingly to one another. Instead, they spent the whole meal ignoring one another while entranced with their smartphones. A "what is the world coming to?!" image has stuck with me (and many others, I might add).

Thirdly, I have been haunted by one aspect of virtual reality which is especially relevant to the prospect of an all-enticing "multiverse". Years ago, I read the popular *Escape from Freedom* by psychologist/philosopher Erich Fromm, and it made a profound impression on me. As a Jew, Fromm had escaped Nazi Germany, and in his *book* he analyzed how people can gradually and unconsciously fall into accepting authoritarian social structures that seem to offer opportunity and security, but are captivating. I now worry that the "metaverse" could become such a trap, where participants happily "play the game" while unconsciously accepting the constraints on freedom to be human.

A book titled *System Error: Where Big Tech Went Wrong and How We Can Reboot* (Reich et al., 2021) looks critically at related trends in technology, including chapters such as "Managing Hackers and Venture Capitalists", "Can Humans Flourish in a World of Smart Machines?", and "Can Democracies Rise to the Challenge of Controlling Technology?" These are key policy issues I cannot adequately deal with in the following chapter, though they surely apply to robotic telepresence and virtual reality, and to the commercial drive pushing these technologies. Will the government entities such as DARPA, NASA, and the FCC, which have had such a large influence in developing the robot and communication technologies enabling robotic telepresence, pay sufficient attention to the challenges described below? Will commercial firms, such as Meta, Microsoft, Google, and Apple, do sufficient self-regulation to prevent widespread harm by the VR/AR technologies and the "metaverse"? It remains to be seen, but both the technical community and lay citizenry can play an important role in contemplating and addressing the challenges.

VIRTUAL VIOLENCE AND ITS TRANSFER TO THE REAL WORLD

For many years, computer gaming technology has attracted especially young people to engage in virtual violence in one form or another. Shooting an assumed enemy with bullets or laser guns, throwing explosives, or driving vehicles at high speeds—all have been standard practice. The "enemy" is typically depicted as evil-looking and shooting or fighting back. Within the commercial market, the number of VR experiences that are peaceful (tours, scenery, education, other passive engagements) is a small fraction of the VR experiences offered. Violence is what appeals to the younger generation, and so it naturally appeals to the profit motive of VR designers.

I have opined (Sheridan, 1993) on the dangerous implications of young people emerging from spending time playing violent virtual games and immediately carrying out the same aggressive behavior in the real world. Many studies have looked at this question, but researchers disagree about whether violent video games increase aggression. A large group of research psychologists believe that the evidence is clear, and that people who play violent video games do have more aggressive thoughts, beliefs, and behaviors than people who don't. They base their conclusions on decades of research in the laboratory and the real world, with a meta-analysis in *Psychological Bulletin* (Anderson et al., 2010).

Opponents of this view find fault with research and point to other studies and meta-analyses that they say show no link between violent media and aggression. The American Psychological Association (APA) weighed in on the debate when it passed *Resolution on Violence in Video Games and Interactive Media*, concluding that the research strongly suggests a link between violent media and aggressive behavior (APA Governing Council, 2015). "Since then, the literature has evolved and, if anything, adds more support to that position", said APA Executive Director for Science Steven J. Breckler, "Nevertheless, this is an area of ongoing research, and other perspectives are emerging". More recently, a new APA resolution on this matter is an update to a 2015 APA resolution, which confirms the link between playing violent video games and aggression.

> But that increase in generalized aggression cannot and should not be extended to link violent games to violent behavior, despite many occasions in which members of the media or policymakers have cited that resolution as evidence that violent video games are the cause of violent behavior, including mass shootings

the APA said. (It is clear that in social science conclusions do not come easily?!)

ADDICTION AND SOCIAL IMPAIRMENT

User addiction and problematic social media use are serious concerns with child computer gamers. This can only get worse in the metaverse being predicted. Internet addiction disorder, social media, and video game addiction can have mental and physical repercussions over a prolonged period of time. These can include depression, anxiety, and various other harms related to having a sedentary lifestyle, *including* an increased risk for obesity and cardiovascular disease (https://en.wikipedia.org/wiki/Internet_addiction_disorder).

Psychological or behavioral dependence on social media platforms can result in significant impairment of an individual's function in various life domains over a prolonged period. This and other relationships between digital media use and mental health have been considerably researched, debated, and discussed among experts in several disciplines and have generated controversy in medical, scientific, and technological communities. Research suggests that women and girls are more easily addicted than boys and men and that it varies according to the social media platform used (https://en.wikipedia.org/wiki/Problematic_social_media_use).

SEXUAL HARASSMENT, BULLYING, PREJUDICE, AND INTIMIDATION

A threat of particular concern in both telerobotics and virtual reality is sexual harassment, which can easily be carried out by an avatar in VR or by a telerobotic device in an actual environment. Virtual reality experiences can lead to simple ethical challenges, to particular kinds of emotions and suffering, or to clear-cut crime. Such crimes include virtual groping and verbal abuse, identity theft, and indecent exposure such as nudity. The potential presence of child predators on metaverse platforms is another concern (Oremus, 2022). Virtual crime, like sexual abuse, and other user safety issues are significant challenges with current social VR platforms and may be similarly prevalent in a metaverse (Basu, 2022). For unruly teenagers (or unruly adults), there is almost a challenge to have an avatar be obnoxious, both verbally and bodily.

In 2022, Keza MacDonald, video games editor of *The Guardian*, criticized the utopianism of technology companies who claim that a metaverse could be a reprieve from worker exploitation, prejudice, and discrimination (MacDonald, 2022). MacDonald stated that she would be more positive towards metaverse development if it was not dominated by companies and "disaster capitalists" trying to figure out a way to make more money as the real world's resources are dwindling.

The prevalent absence of law and order in VR experiences is likely to provoke actual disorder in real life. One might expect that a female VR player who is groped virtually feels the same as being groped in real life. Bodysuit technology may develop to a point where wearers in VR could actually feel unwanted touch or punches. A key consideration is that since being murdered or raped in virtual space is not objectively possible, the very absence of actual consequences is likely to result in more people engaging in those acts in VR.

The metaverse may magnify the social impacts of online echo chambers and digitally alienating spaces. It may abuse common social media engagement strategies to manipulate users with biased content (Shou, 2021).

INJURY AND PHYSICAL SAFETY

Early industrial telerobots were particularly unsafe because they used hydraulic actuation. But even electric motors are powerful enough to injure a nearby person if there is a sudden glitch in the control loop. Happily, some newer robots are made with a softer touch, even adjustable stiffness of the endpoint or hand, as described in Chapter 2. Looking to future applications, there is a need that the operator to be aware of potentially harmful forces and movements. Better kinesthesis and tactile sensing will help.

The Occupational Safety and Health Administration has categorized four primary types of workplace accidents that can involve robotics or automation.

1. **Impact or collision accidents:** The workplace might soon be crowded with robotic arms, autonomous conveyor belts, and wire-driven delivery systems. Unfortunately, the likelihood of being struck by a piece of track machinery has greatly increased.
2. **Crushing and trapping accidents:** As an added layer of danger to an impact accident, workers might become trapped between a robotic arm and another piece of equipment.

3. **Mechanical part accidents:** If the autonomous equipment breaks down or malfunctions it could cause an accident. A gripper mechanism could fail, dropping materials and injuring workers.
4. **Other accidents:** These include shock injuries, burn injuries, and the release of pressurized fluids—these malfunctions can cause critical injuries.

VR has also been associated with physical injuries. Carpal tunnel syndrome, stiff shoulders, eye-strain headaches—these are all well-known side effects of prolonged computer use. But what happens when you step away from the desktop and into virtual reality? A recent study assessed how some common virtual reality movements contribute to muscle strain and discomfort. There are efforts to ensure future user safety in this fast-growing technology that is used not only for gaming but also increasingly for education and industrial training (Rosbach, 2020).

Using telepresence robots in the school setting opens up the possibility of potentially inappropriate contact between a child and the robot (or robot operator). Inappropriate contact is legally recognized as a component of the tort of *battery*. What is considered offensive contact, however, hinges upon violation of "a reasonable sense of personal dignity". Thus, an individual touching the robot in an attempt to get the robot's attention may not qualify as offensive contact, whereas kissing the robot or striking the robot may "offend the ordinary person" and qualify as *offensive contact.*

PRIVACY, SPYING, AND THEFT OF PERSONAL DATA

With respect to telepresence, privacy is a major concern. Consider the situation with aerial drones. Anyone with an inexpensive aerial drone can invade the private space of someone else, do mischief, and not be detected. A drone can easily sidle up to the window of a house or commercial building, have its GPS set to accurate hover, set its proximity sensors to hold a few feet from the window, and have its camera turned to look directly inside.

Telemedicine, which we normally regard positively, is particularly susceptible to privacy infractions. Reindl et al. (2021) examine the legal implications of telemedicine and make the following points in regard to that technology: (1) VR and AR legislation must always be anticipative of innovation. The visceral nature of virtual/augmented reality will present notable roadblocks and question the fundamental norms of freedom and harm in the real and virtual world. Researchers in the European Union have commented on a lack of an international standard for diagnostics data transmission and liability in telemedical systems, which hinders pushing telemedicine technology forward. (2) Diagnostics need to be integrated to ensure physically safe, unified, and data-protected haptic telepresence. (3) Voluntary standards should identify and classify threats, identify anomalies, and suggest response actions. Presumably this would simplify integrating components from different providers. Companies could use such standards to prove that their products or services comply with the technical requirements of the relevant EU law. (4) There should be some mandatory technical information available that specifies system operating characteristics and qualities used in system evaluation. (5) Legislation should clarify the levels of quality one expects and how they are supposed to be tested.

Privacy has always been a major concern with respect to VR. VR gaming and training platforms process and store much personal data of users to further their advertising and business models. For instance, participant reaction times can be used as health data and sold to interested entities. Users may also be asked to provide data about their location, age, race, preferences, etc., which can be used in their identification in real life. Unfortunately, privacy regulations today have not caught up with privacy infringement in VR.

Concerning the metaverse, companies will likely collect users' personal information through interactions and biometric data from wearable virtual reality devices. Meta is allegedly planning on employing targeted advertising within their metaverse, raising further worries related to the spread of misinformation and loss of personal privacy (Jackson, 2022). It has been argued (Intel, 2021) that the amount of data collection in the metaverse would be greater even than that on the current Internet: "If you think about the amount of data a company could collect on the World Wide Web right now, compared to what it could collect with the metaverse, there is just no comparison" (https://www.intel.com/content/www/us/en/newsroom/opinion/powering-metaverse.html)

Remarkably, privacy invasion in virtual reality seems fully recognized by the VR research community. Tracking of emotions, expressions, and physical behavior in VR goes well beyond what is available to Internet sales firms like Amazon that simply track purchase preferences. Elected officials are already asking questions about how technology such as the Oculus headset allows personal data to be acquired. Users of such devices are typically asked to give consent to terms and conditions before using such devices, and they perfunctorily do so, giving companies legitimate rights to collect, share, and use such data. Relevant statutes relate to wiretaps, public accessibility, manufacturer's ability to intercept data, user access to stored communications, law enforcement access, owner obligations, and state laws for access and recording.

Because of their ability to sense (i.e., see and hear) remotely, and to transmit information about a given physical setting to a faraway user, telepresence robots necessarily raise legal concerns about surveillance and the expectations of those with whom the robots interact. These concerns may be particularly acute in the workplace, given commercial sensitivities around materials discussed or shared in office settings. In the United States, federal law prohibits the interception of, or eavesdropping on, electronic communications (the Wiretap Act), as well as accessing stored electronic communications without authorization (the Stored Communications Act). Listening in on a telepresence robot while it is attending a meeting might be a violation of the Wiretap Act (Barabas et al., 2015).

With respect to privacy regulation, a proactive approach by governments of various jurisdictions could be instrumental in setting the limits on the type of personal data that can be collected, the extent of the data, and the ways in which it can be used. A wider interpretation of informational privacy is also necessary for judicial systems, which will facilitate further privacy regulations.

HACKING AND CYBERSECURITY

While telepresence gains have been rapid and promising for many new applications, this technological capability is certainly amenable to hacking. A hacker could break into the communication channel and take over control of the telerobot, or make some less dramatic hack by inserting false information into the sensory feedback. More seriously, for instance, if telepresence technology is used to make critical adjustments to a nuclear power plant, water supply, or other municipal installation that affects thousands of people, small hacks can have very broad effects.

The "internet-of-things" almost invites hacking. It is not clear how great is the need for encrypted communication, but there is plenty of room for deception and interference in the communications, resulting in serious loss of control (Yacoub et al., 2022).

Virtual environments demand that the software underlying the whole system be dependable. For example, when a teacher using VR instruction is isolated from a class of children, she cannot as easily see and monitor the result of computer hacks *that mischievous students* might regard as good fun. Virtual conferences lack the opportunity for handshakes and reading of body language, so agreements or disagreements may not be as evident to participants, and computer hacks might show bodies moving differently than how they are actually moving. One can imagine uninvited avatars controlled by hackers engaging in all sorts of offensive behavior.

Who authorized this action? Who is in control here? Who set these rules? Who are you, Mr. Avatar? Clearly both telepresence technology and virtual reality technology provide the opportunity for anonymity and intentional false identity with respect to who is controlling the telerobot and who is behind a particular avatar in a virtual environment. Anonymity poses many legal questions (Lanxon, 2021; Antin, 2020).

EMPLOYMENT AND MARKET DEVELOPMENT

A 2021 MIT report called *The Future of Work* claimed that the major near-term problem in the world of work is the disparity between jobs held by the educated and those held by the uneducated. Ability to understand and use technology is part of this. Artificial intelligence (AI) and robotics will gradually come to dominate many jobs, and will not so much put people out of work, but will require technical skills to operate and manage the technology. Eventually I do believe robotic telepresence coupled with automation will create more jobs, especially to build infrastructure, initially in developed countries, but gradually in less developed countries. The job disparity cited above will place a major demand on education, and innovations in VR will have a large effect on how education can be utilized in massive open online courses (MOOCs).

The very large tech firms, Google, Microsoft, Apple, Amazon, and Meta are all gearing up to produce VR equipment such as head-mounted displays (HMDs) and the software that enables game playing, education, virtual travel, and new forms of social media. A piece in *The New York Times* (Metz, 2021) was headlined "Power brokers of big tech see piles of money to control access to the metaverse". The article admits that to date VR has not lived up to the hype that some people provided, but predicts that this will change. It cites one research firm that claims the market for "metaverse technologies" (technical devices and services) topped 49 billion in 2020, and will grow by more than 40% each year. Perhaps the best-known deal was Facebook's 2014 acquisition of Oculus, which was already the leader in selling head-mounted displays, but Facebook spent several years and lots of development money to make it the current most popular HMD version. Apple is reportedly making a ski-goggle-sized HMD. Google had produced a product called Google Glass, which flopped on the market, but is reported to have a new goggle-type HMD coming along. Intel produced a prototype glasses product called Vaunt, but then sold the patents to North, a start-up acquired by Google. We don't yet know how well the Microsoft HoloLens will fare. But clearly, there is dramatic stirring within the VR tech world, mostly on HMDs.

In a 2022 article for *The New York Times*, Lauren Jackson argued that the metaverse is "stalled from achieving scale by a lack of infrastructure for both hardware and software, a monopolistic approach to platform development, and a lack of clear governance standards". Though others such as Nick Bostrom (Wikipedia) have argued that future technological developments, such as "more realistic computer graphics" and improvements in artificial intelligence, will continue to incentivize user engagement, which will lead to the "normalization" of the metaverse ("Immersive Web Working Group". *www.w3.org*. Retrieved 2022-03-01).

Raja Koduri, senior vice president of Intel, said that a truly persistent and immersive computing, at scale and accessible by billions of humans in real time, will require a 1,000-times increase in computational efficiency from today's state of the art (Koduri, 2021).

The World Wide Web Consortium (W3C) is an international community where member organizations, a full-time staff, and the public work together to develop Web standards. Led by Web inventor and Director Tim Berners-Lee and CEO Jeffrey Jaffe, W3C's mission is to lead the Web to its full potential. Within W3C, there is an Immersive Web Working Group devoted to bringing high-performance virtual reality (VR) and augmented reality (AR) (collectively known as XR) to the open Web to interact with XR devices and sensors in browsers (www.W3C.org).

Any attempt to regulate VR can have unintended consequences. For example, any legislation that is aggressively protective can stifle technological progress. Some believe that for now, self-regulation by VR companies may be the best solution currently available. Self-regulation allows VR content providers to develop industry-friendly policies and participate actively in the lawmaking process. This obliges companies to take proactive steps in formulating legally and ethically sound strategies for dealing with infringement. In contrast, overarching regulation has the possibility of undermining the attractiveness of VR. The extent of possible tort violations is large, so a "wait-and-see" approach might make more sense and be accompanied by improvement on a case-by-case basis. Where the technology is heading is unpredictable. In the same way that gene editing technology has produced much discussion among biological scientists, so, in the VR and robotic telepresence world, engineers and government officials need to deliberate as to appropriate needs and limits on regulation.

LUDDISM AND ORGANIZED REVOLT

The Luddites were members of various bands of English workers who, between 1811 and 1816, destroyed machinery, especially in cotton and woolen mills, which they believed was threatening their jobs. They identified with a legendary "King Ludd", who supposedly championed their cause. The term has come to apply to people who are opposed to new technology, seeing it as threatening, and are even out to destroy its development.

We could soon see examples of such opposition to the VR metaverse or to robotic telepresence. One example is the vote taken by a group of 4,000 anonymous Google employees who opposed the company's work on a Pentagon project using artificial intelligence (AI), which could be used to improve military drone targeting. Due to the employees' actions, which included an internal petition to company management, Google ended its work on Project Maven when the contract expired and announced that it would focus on "socially beneficial" AI and avoid work that causes "overall harm" (https://www.nytimes.com › google-pentagon-project-maven).

It is not unthinkable that military application of telepresence technology could elicit other protests. Similar protests are likely using VR. For example, *Liberate* Hong Kong is a 3D single-player simulation video *game* developed by students during the 2019–2020 *Hong Kong protests* (https://en.wikipedia.org › wiki › Liberate HongKong). These were young people who hoped this would allow citizens to experience their anti-government demonstrations from the view of a "frontline protester". The short but detailed first-person game allows players to dodge tear gas, duck behind burning barriers, and run from riot police. "It allows you to experience what crisis the frontline protester might have been through", said the 30-year developer, who covered her face and did not want to give her full name because she had participated in the often-illegal protests (https://en.wikipedia.org› wiki Liberate Hong Kong).

ALGORITHMIC BIAS, AI BIAS, AND SOFTWARE LIMITATIONS

Both robotic telepresence and virtual reality will make increasing use of computer decision-making and control. Both are complex based on software and hardware that may or may not have been proof-tested in the context of present use. Some errors, once caught, are easy to fix. Others, such as misidentification of faces, especially for dark skin tones, are not easy to fix (Girdhar, 2020).

Decisions are often based on a lack of adequate evidence. They may be based on models and algorithms that are out of date, are misapplied to the current situation, or are beyond the assumptions upon which they were designed. Artificial intelligence based on machine learning by neural nets is necessarily predicated on the limited data set on which it was trained, and this can easily result in bias (Misek, 2020).

MIT's Prof Norbert Wiener, known as the "father of cybernetics", worried a lot about this. His last and most popular book, published in 1964, called *God and Golem Incorporated*, issued a warning by comparing the computer to the Golem of Hebrew tradition, a monster that did not always do what people expected. The computer does not always do what the programmer intends, and this can be catastrophic if the computer is given great responsibility (Wiener, 1964).

INTELLECTUAL PROPERTY RIGHTS

Intellectual property in VR is a major concern if one is the owner of said intellectual property (IP). Intellectual property issues include direct infringement, copyright, contributory and vicarious infringement, and communication or power service provider responsibilities

For example, in the VR game called "Second Life" (Wikipedia), players can visit various public places and use merchandise that could violate trademarks and copyrights of established real-world brands. As they now exist, VR systems do not prevent users from importing photographs, music, brand names, and other IP-protected material into their virtual environments—without bothering to obtain required permissions from the IP owner. Also, the fact that VR users log in from different countries, whose IP laws may differ from one another, poses jurisdictional problems. The possibilities for intellectual property infringement are many.

There may be consequences for such infringement. A person might distribute copies of a painting in the virtual world, which is copyrighted by the painter in the real world; if so, such a person can be held liable for copyright infringement. If a virtual experience by a VR platform is created which uses real-world intellectual property that is copyrighted, the owner of the IP may hold the VR platform liable. For example, in one VR infringement, ZeniMax Media sued Oculus VR for using ZeniMax's prototypes that lowered the latency between a user moving their head and the VR display updating (ZeniMax v Oculus, *Wikipedia*).

The above suggests a need for IP laws to clarify the obligations VR designers must consider to conform to users' IP rights. For example, traditional IP laws require permission from the owner regarding applicable rights. The use of trademarks "in commerce" is an essential component of trademark infringement. However, using a logo in VR may escape liability by claiming a "no commerce" exception under existing trademark laws. The same is true with the "fair use" exception under copyright laws of several jurisdictions.

In contracts with VR content creators, IP holders should state specifically the ownership and liability arising in case of breach/unfair use of the intellectual property. Also, the development of VR will likely see more cases of violation of "fair use standard". In any case, it would help if corporations and governments come together to formulate industry-specific standards for fair use (Howard, 2018).

It is generally good practice to put IP and related contracts in writing to minimize ambiguity. When there is sufficient evidence of oral contracts, such as in the form of witnesses or video recordings, oral contracts may be enforceable. Otherwise, there is a problem. Telepresence may mask non-verbal cues, or render some of them impracticable. Telepresence may not fully convey a party's tone of voice, facial expression, body language, and/or gesture that can contain contextual information important for the interpretation of a contract's terms.

CONCLUSION TO CHAPTER 5

The rapid increase of both hardware and software technology for application to robotic telepresence and virtual reality raises many critical questions. Academic and industrial researchers in these areas recognize the challenges, but governments and the lay public are slow to understand what is happening. Relevant professional societies with global reaches, such as the Institute of Electrical and Electronics Engineers (IEEE), or national scientific deliberative organizations, such as the US National Research Council, have the capability to bring together technologists, social scientists, and policy analysts to better articulate the issues and make recommendations. I urge them to do so.

REFERENCES

Anderson C, Shibuya A, Akiko S et al.: Violent video game effects on aggression, empathy, and prosocial behavior in Eastern and Western countries: A meta-analytic review, *Psychological Bulletin* 136(2): 151–173, 2010.

Antin D: How avatars support a private and unique metaverse. https://superjumpmagazine.com/how-avatars-support-a-private-and-unique-metaverse-b.

APA Governing Council: *Resolution on Violence in Video Games and Interactive Media*, Washington, DC, American Psychological Association, 2015.

Barabas C, Bavitz C, Matias J, et al.: Legal and ethical issues in the use of telepresence robots: Best practices and toolkit, Working Draft: March 27, 2015, presented to We Robot 2015 Fourth Annual Conference on Robotics, Law & Policy April 10–11, University of Washington School of Law, 2015.

Basu T: The metaverse has a groping problem already, *MIT Technology Review*, January 19, 2022.

Bostrom N: https://en.wikipedia.org/wiki/Nick_Bostrom.

Girdhar A: The limitations of virtual reality, 2020. https://www.appypie.com/virtual-reality-limitations.

Howard B: Protecting intellectual property rights in the billion-dollar world of virtual gaming. https://techcrunch.com/2018/01/23/protecting-intellectual-property-rights-in-the-billion-dollar-world-of-virtual-gaming/.

https://en.wikipedia.org/wiki/Internet_addiction_disorder.

https://en.wikipedia.org/wiki/Problematic_social_media_use.

https://www.intel.com/content/www/us/en/newsroom/opinion/powering-metaverse.html.

https://www.nytimes.com › google-pentagon-project-maven.

https://en.wikipedia.org › wiki › Liberate HongKong.

"Immersive Web Working Group". www.w3.org, 2022.

Jackson L: Is the metaverse just marketing? *New York Times*, February 11, 2022.

Koduri N: Powering the metaverse, *Intel Newsroom*, December 14, 2021.

Lanxon N: Welcome to the metaverse: What it is, who's behind it, and why it matters, *Bloomberg*, September 29, 2021.

MacDonald K: I've seen the metaverse, and I don't want it, *Guardian*, January 25, 2022.

Metz C: Everybody into the metaverse: Virtual reality beckons big tech. https://www.nytimes.com/2021/12/30/technology/metaverse-virtual-reality-big-tech.html.

Misek R: Real-time virtual reality and the limits of immersion, *Screen* 61(4) Winter: 615–624, 2020. https://doi.org/10.1093/screen/hjaa058.

Oremus W: Kids are flocking to Facebook's 'metaverse'. Experts worry predators will follow, *Washington Post*, February 7, 2022.

Reich R, Mehran S, and Weinstein J: *System Error*, New York, Harper-Collins, 2021.

Reindl A, Rudigkeit N, Ebers M, et al.: Legal and technical considerations on unified, safe and data-protected haptic telepresence in healthcare, in *Proceedings of the 2021 IEEE International Conference on Intelligence and Safety for Robotics*. Nagoya, Japan, March 4–6, 2021.

Rosbach M: *ScienceDaily*, Oregon State University, January 8, 2020.

Sheridan T: My anxieties about virtual environments, *Presence: Teleoperators and Virtual Environments* 2(2) Spring: 141–142, 1993.

Shou D: I want my daughter to live in a better metaverse, *Wired*. ISSN 1059-1028, 2021.

Wiener N: *God and Golem Incorporated*, Cambridge, MA, MIT Press, 1964.

Yacoub J-P, Noura H, et al.: Robotics cyber security, vulnerabilities, attacks, countermeasures, and recommendations, *International Journal of Information Security* 21: 115–158, 2022.

ZeniMax v Oculus, *Wikipedia* (civil lawsuit)

Appendix 1: Responses to Polling

Here is an email message I sent to: (1) a few of my children and their professional associates, which I am calling a "younger cohort", and (2) a small number of contemporary friends in the community where I live, which I am calling an "older cohort". The aim was to see what the immediate reaction would be to a futuristic image of telepresence, a concept more or less new to both. Then I provide their unedited responses in no particular order.

Friends,

I am writing a book, called *Telepresence: Actual and Virtual.* I want to get reactions, positive or negative, or both, to the following scenario. I plan to include the polled reactions in my book. Thanks for your help.

Imagine yourself in a future world where you are controlling a robot to perform some job. The robot is remote, an arbitrary distance away, a task location of your choosing. It is mobile, you can move it anywhere. It can walk or crawl, fly like a bird, or swim like a fish. It can be of any size, big like a truck, or small like an insect, appropriate to the required task. It may not conform to your idea of what a robot looks like. It has multiple tiny camera eyes, steadily peering in all directions. It has some form of robotic arms and hands, well endowed with force and touch sensors, and capable of delicate or forceful manipulations. It can listen and locate sound sources, and even smell.

As its controller, you can make it do a variety of tasks, such as might be found in a kitchen, on a farm, in a factory, on a construction job, or in a hazardous chemical, nuclear, or fire-fighting environment. It can tend to a baby or an elderly person in need of a hand. Or it can just attend a meeting with other such robots. Any of these tasks can be partially automated, so that the robot does not need your undivided attention, and so that you can intermittently instruct it to do pieces of the task, while you check on progress at your convenience, and learn what it did, saw, or heard.

If you, the human operator, wear a head-mounted display, you can turn your head in any direction and see what the remote robot sees. If your hands and arms are placed on mechanical arms that are the same as those of the robot, the robot's actions can duplicate your own, and you can feel what the robot feels. In this way, you will have a full sense of presence at the remote environment. This is called *telepresence*, and it is well demonstrated. The robot is an avatar, a remote you. You will be able to do many things remotely, and feel *telepresent* in that remote environment.

None of the above is science fiction. It is all under current development, some of the pieces further along than others, already used in space and undersea research, manufacturing, surgery, and virtual reality gaming. Gradually, not next year but over coming decades, the speed, accuracy, power, and cost will become more agreeable, and such technology will become more widely available for anyone to use.

In a few words, considering both the positive and the negative potential, what is your gut reaction to a future of robotic telepresence?

RESPONSES OF THE YOUNGER COHORT

JB

I think it has amazing potential to reduce potential human harm and improve overall human efficiency. I am especially hopeful that this augmentation paired with real world interaction keeps human interaction still based in the real world and not fully digital. My fear is that progression then draws humans into using similar sensory technology to migrate into a fully digital world where humans will no longer interact with one another in a meaningful way in physical space. This is not a future I would enjoy. However, I would enjoy a hybrid physical/digital world paired with human technological augmentation.

JB friend 1

I'm certainly excited about the prospect that you're able to do more work that wasn't able to be done before. In many ways, having telepresence is almost better than having only artificial intelligence doing your work because at least you are in the "loop". I think it's important for those who are doing these telepresence roles to be prepared for what it entails—being a part of the loop. But far enough away can still numb us to the realities of what is going on, like drone strikes. So it's important to have strong mental coaching.

JB friend 2

My gut reaction is yes, this future is already here. In my time at MIT I had the opportunity to meet the first-hand scientists behind many of these innovations. Like with any new technology, I fear the negative implications it can have, especially in areas such as invasion of privacy and warfare. Of course, there are benefits to be had, for example, around the augmentation of abilities for people with disabilities, or care for those who are dependent on caregivers. The balance will lie in adequate regulation that considers the potential for harm and protects individuals over the interests of states and corporations. In summary, my gut reaction is of awe at what some of this technology can do (the field of robotics is fascinating to me), and also a sense of caution and responsibility for the ethical implications of any of these technologies that enable telepresence and could be used for the wrong purpose.

CB

There are some potential great benefits to technologies like this (the list is exhaustive and these are just some ideas):

- In medicine
    - better, more accurate, more life-like prosthetics and restorative care
    - precision surgeries on a minute level

telehealth visits become the same as being in person
highly accurate remote monitoring
- In exploration
- In making hazardous jobs more safe

However, I think there are also many potential negative consequences. Again the list is long and here are just some:

- A further divide in human communication as we know it—face to face.
- Ethical dilemmas around artificial intelligence and what constitutes life.
- Legal disputes over ownership of the machine. What happens if the machine causes damage to something or someone? Who is responsible (the "driver" of the machine, the company that created the machine, the software developer that created the AI)?
- Issues of security and privacy.

Potential to make people less present, which potentially dissociates you from the situation at hand (i.e., do something you wouldn't do in person but feel okay doing over telepresence)

- Cause a loss in a sense of reality.
- Further disparities in rich vs poor (especially at the beginning).
- I think, in general, telepresence can be an incredibly powerful technology that yields great outcomes. But I think it is also potentially very dangerous and we as a society need to be cautious with the implementation of this type of technology. An interesting conversation to be had is who are the people, who are going to decide the path of this technology, and who will this technology impact the most (and how)?

I may be coming to you with more questions than answers. This is an incredibly interesting question and scenario. I think in short I would say: proceed with this technology, but with great caution.

RS
That future is coming and we need to adapt. We can learn a lot about the communication revolution which is slightly ahead of the presence revolution. Digital communication has been a boon to productivity and has enormous positive implications. It has also created a lot of headaches because of anonymity, a deluge of information making it difficult to sort and difficult to determine truth.

Similar challenges will occur because of a revolution in presence. We will have to develop ways to identify agency because people will be prone to use their robots to rob banks or hurt other people. I can imagine a whole travel industry where people will travel vicariously through their robots without danger or inconvenience. And there will be a giant challenge to adapting in time.

As the speed of change accelerates, our greatest social challenge will be keeping up with the social and legal adaptations to new technologies. Something like virtual presence will not only pose huge challenges but make other giant changes possible which will, themselves, require adaptations. Ultimately, this might lead to the demise of our species. Destroyed by our own fabulous successes.

SB
My reaction is negative. I feel that it would negatively impact my interactions in the world because I would not be having the experience that the robot would. There are situations where its use could be both positive and negative.

MB
My gut reaction is fear that development happens faster than analysis of the social, moral, and ethical ramifications.

NS
Our family discussed this at dinner tonight.

Here are the preliminary thoughts:

Concerns include moral values compromised by the fact that robots could create crimes:

- Once we begin this process, it will be very hard to draw the line and return to a previous methodology (not using AI).
- There is no oversight and accountability.
- Jobs would be replaced, supplanting other workers who will then need other jobs.
- Equity issue—if this is expensive, only the rich will have this extraordinary capability creating more division among the population.

NS2
When it comes to professional opportunities, I'm all in. Anything that would allow me to cut down commuting time and do the work from home.

My biggest hesitation about this kind of tech is that it's a one-way experience. Maybe I could experience taking care of my baby, but she wouldn't experience me taking care of her, and so I wouldn't want to have a barrier between us.

KB
My gut says this is amazing and inevitable, but there are risks/dangers.

On the positive side: As you know, there is great potential for use of this technology. Imagine a small robot that could be swallowed. One could perform surgery on the small intestine with no need to create large wounds in the abdomen to access the site.

On the negative side: Does the "depersonalization" of the robot create issues? As you know, when humans use some machines, it can provoke or promote behavior that one would not engage in if in person. An example is road rage where one uses the car to threaten another when that threat might not occur if there was a personal interaction. Another example might be how much easier it is to use a military drone to cause harm as compared to causing harm through your direct action against another person.

I am very interested to hear what others have to say.

OS
I am apprehensive. I try to be "present" to pay attention to and appreciate my surroundings. Will it be harder to be "present" if I can be "telepresent" anywhere? Also, what will telepresence mean for the demands placed on me as a worker? It's hard enough to be expected to be in one place at one time.

SP
All the different factors—dimensionalities, senses/sensations, environments, chronology—that can be affected make this seem like a more complicated question than it truly is. But as you point out, most of this stuff is operating now, if only in primitive form. Waldos have been in use for decades. An entire generation has grown up accustomed to the idea of playing in virtual reality, which pretty much exactly fits the description you give of working through a robot ("remote, an arbitrary distance away, a task location of your choosing. It is mobile, you can move it anywhere. It can walk or crawl, fly like a bird, or swim like a fish. It can be of any size, big like a truck, or small like an insect"). Of course, drone operators are already "working" in that kind of reality.

In short, people have been adapting to—or perhaps it would be better to say "acquiring" —telepresence for decades, if not centuries. It really depends on how you want to define matters. Isn't radar/sonar basically a robotic extension of our sense of hearing, in a kinesthetic fusion of sight and sound? What is a telephone conversation? For that matter, what is an exchange of telegrams? It might seem like I'm pushing toward a *reductio*, but there's a point to my prodding, which is that I think there is a line we will cross which will have profound consequences for consciousness. That line is where our proprioception becomes indistinguishable, or distinguishable only with effort, from our telepresence. Where the interface of robotic and organic becomes—transparent? seamless? I don't even know what word applies. What is my gut reaction to this? To be honest, I can't really say. I know that if human beings survive long enough that this line is crossed, the consequences will probably be both much greater and much less than I think, which kind of makes any feeling I have about it besides the point.

RESPONSES OF THE OLDER COHORT

JM
My primary reaction is one concern: The depersonalization that can occur with the substitution of robotic activity for human intervention. A wide variety of other feelings accompany this technology depending on how it is being used, but it is our potential for separating ourselves from the consequences of our actions, especially as they apply to others, that worries me the most.

FH
The benefits will be wonderful.

The risks, however, are huge—we humans carry potential aggressiveness in our DNA and many of our cultures. How could the world be protected from the controllers' aggressive use of the robots that you describe? Otherwise, they would multiply manyfold what gun prevalence inflicts in our country today.

CH

I have mixed feelings about this. In general, it's the kind of progress that should be supported, as it has considerable reasons to be good for society and the world in the long run. Change is always difficult and upsetting, I hear the innovators say.

However, and this is a big reservation, I see a couple of problems:

1. The more jobs this telepresence creature is asked/tried to do, the more unemployment will result. They are likely to be made to replace current jobs, rather than doing a yet unidentified role. The social cost of unemployment needs to be considered, and somehow controlled by regulation or legislation (which in this climate seems nearly impossible). Unchecked, I see this technology as providing a real service to the richest of the rich—the top 1% and injuring the bottom 50% or more. Massive civil unrest could result.
2. But the technological answer to that would be to create a number of these things that could defend the lair of the plutocrat-owner physically. Is there some reason these couldn't be developed and trained as storm-troopers, able to do all the physical tasks that troops or security details could do?
3. Then there is the question of whether these things could be able to think, even rudimentary thoughts, and then self-program themselves to make or do the answer, even if these were not among the initial commands. If I were one of these, it seems to me to be one of the closest sci-if worries to conclude that such-and-such could be done if there were just 100 more of me, and then make them. I suppose one could put blocks in the algorithm that they use to think on—if the creator really wanted to prevent that.

BW

Bring it on! A few decades ago, I spent two years doing (on-the-spot) research in the Bibliothèque nationale in Paris. The environment included a chilly (although handsome) hall, mostly efficient but sometimes grumpy staff, a hard chair, and the pallid northern European winter light. Stopping for lunch meant losing nearly an hour's work-time. Compare that to sitting in my favorite chair, at my well-lit desk, steps away from the kitchen. Type "BNF" and if it doesn't have what I want, have a look at the catalogs of the British Library or Houghton or … And note-taking? Press *print*, or *save*. Why not both! And the rest of the time is free for the joy of reading, knowing I can always return to the precise text. I suppose I've become my own (happy) robot.

And it's not bad news for people without a lot of money. Every public library offers Internet accessibility (and help). The whole computer–Internet world has become so accessible, so quickly, that what were unimaginable riches of information that we can get up from our chair, go to the information source, and sit down again.

My first computer was an IBM PC1—the first American PC that could deal with foreign language characters. (I was writing a book about the history of French cooking. Can you imagine such a book without accents!?)

EB
I fear too many robotic avatars for mundane stuff like getting the paper, a cup of coffee, etc., will encourage folks to sit on their duffs too much and miss out on the needed movement of their bodies to stay healthy??? Likewise, not having "hands on" when the avatar won't improve the function could wind up being like Zoom, which is great but not as fine as being somewhere participating in person!!!

SB
I think using such robots to deal with biohazards, fighting fires, etc., is of huge benefit. Using it for many routine tasks like filling Amazon orders in warehouses seems quite safe. There is the human cost, however, when these human jobs are replaced by robots. If a human is paying attention when the robot does more complicated tasks with potential risks, I am OK with that idea in theory. However, caring for an infant or elderly person by a robot is abhorrent. Human touch is essential at every age, but most especially for children and seniors.

PG
Good luck to my grandkids! Some are already there.

JH
The potential—both positive and negative—is so great that the prospect of such a potential widely available seems rather overwhelming. I guess I think of the impact of the Internet and of social media in our world—how many wonderful but also really terrible consequences they have had.

MH
My gut reaction is that it will be able to do everything … cook meals, babysit, change diapers, clean my house. It'll be my slave … until … it becomes my master. Is that still science fiction?

AM
Your question reminds me of Jules Verne, the famous nineteenth-century novelist that wrote *Voyage to the Center of the Earth*, *Around the World in 80 Days*, *Twenty Thousand Leagues under the Sea*, *Voyage to the Moon,* and many other books trying to look into a future he knew nothing about. Then he wrote Paris, XXI century and tried to predict what Paris would be like in the XXI century … He was totally wrong about absolutely everything … could not predict the automobile and could not predict public transportation … there is always a big unknown but the question is always interesting and gets in the way of predictions into the future.

JM
Boy, is your list of responses going to be interesting!!!!

I'm not a "techie" of any kind, for sure, but I certainly use and am most grateful for all the new advances so far!! The one, really serious caveat, I have for this future scenario, is the continued diminution of human contact!

I have often witnessed, now for example, sitting in a restaurant and watching a table of four to six young people all together and every single one on his/her cell phone!! Little, or no, interactive conversation!! We wonder if kids are reading books?

We are so grateful for FT and Zoom in this C-19 era, but now worry that too many businesses and families will remain remote!! Are parents going to put their kids with other kids, in person, or get them involved in all sorts of virtual stuff? What are schools going to do?

If working robots is the wave of the future—especially for the daily round of activities, I see humans gradually becoming amoebas with a pseudopod to push buttons!!!

Or, in what you describe, with a head-like protuberance that manages the robot emerge. Well beyond scary!!!

Will the diverse and active "brains" behind all this future accede to any kind of controls?? I was supposed to be brief!! Sorry! In a few words, my gut reaction is fear of increasingly centralized control without recourse. Do we all get to see the results of your polling?? Fascinating!!

JH

Given that teleoperation technology is already here, I'm sure it will be used. We already have limited examples like drones and robotic surgery. Some attractive uses include working in hazardous environments like cleaning up a melted-down reactor, or mining, or industrial processes involving dangerous materials. The huge draglines and trucks already used in open pit mining already feel like robots, with the operator somewhat distant from what's actually happening. Telepresence just makes the human interface more intuitive. Imagine a crane fitted with cameras and sensors at the pick-up end, so the operator feels like he is flying around and picking objects to be moved.

Scaling down is also attractive, for example, in manufacturing or repairing small objects and in surgery. I've often fantasized about being able to see the back of my head or to zoom in on and remove a splinter. There are pills you swallow for a visual tour of your gut, and fiber-optic cameras that can be run through arteries. What if they could be run and moved around without a physical connection to the operator? There are other physically inaccessible environments that could use telepresence: space, the bottom of the ocean, etc.

On the downside, I worry about war robots. We are already seeing the moral dilemmas with drones attacking the wrong people, but even before drones, pilots bombed people they couldn't see, often killing civilians. If I step into my office next door, "put on" a warrior telerobot, and rampage around some remote village, what have I become? What if telerobots become so realistic that I can pretend to be a different person, evading liability or even causing an innocent third party to be blamed.

Teleoperation helps to enable full automation. Every input and action in a teleoperation system can be recorded. Once recorded it can be replayed without the human in the loop. More sophisticated analysis leads to more flexible responses. Use AI to recognize patterns and select scenarios, and you may have the potential to replace almost any human activity, benign or malign. Scary.

JS
It would depend as to whether I am on the end that the robot is doing something to or for or whether I am the one controlling the robot to accomplish a task which I help guide. I would love if something like that could pick up heavy objects that I no longer can do, or dig holes in the garden for me to plop a plant into, or carry the paddle boards up and down the hill to the water——If it can open cans, bottles, etc.—it could be my best friend and I might even get quite fond of it. Since it is a technical wizard, I suspect that I will have trouble being as clever as it is and like my computer. There may be moments when I'd like to toss it out the window if it is not doing something that I thought I told it to do. The ratio of trouble running it to the benefit that it provides is crucial. If it keeps track of me and sends that info on to some central agency to sell my info would be the tangle—who wants to be followed all the time——anyway it's an initial thought—I'll think some more.

EB2
I simply would hope that robots would be used only for gaming and specific uses when a robot would enhance a person's skill vs replace what can naturally and more conveniently be accomplished by a person, i.e., robotic surgery, which already exists.

I'm concerned that we currently are over-digitized, e.g., how will we flush some toilets if the power fails? Just because something can be done with a robot doesn't mean that it should be implemented—such that a person will not be able to easily step in, or override, if/when the robot has a problem.

Complex digital systems will always have bugs, much as humans will occasionally make mistakes. Special situations when the robot may malfunction need to be given careful consideration. Robot/human operations ideally should result in improved outcomes for both. Not sure I've answered your question … I have both positive and negative reactions to robots. If self-driving cars are "robotic", I am looking forward to when an acceptable human/robot interaction is developed.

BG
My gut reaction is awe. I think this kind of robot will be as commonplace to our great- and great-great-grandchildren as computers are to our grandchildren. A little more specificity would be helpful. I'm trying to use my imagination about a robot's usefulness in Brookhaven or in a home where elderly residents often need intimate assistance. Could a robot clean an elderly person who has soiled himself? Could a robot help a person get in and out of a shower or a bed? I'm pretty sure that other tasks are more easily accomplished by a robot, i.e., cleaning up spilled liquids and foods, cleaning up broken glass, and dispensing medication. I would imagine that a robot could even give a massage. How about going to a store to buy groceries?

It's a fascinating subject. Intimate care by a robot would eliminate many of the difficulties inherent in trying to find and keep competent caring home health aides.

The negative side, of course, is the lack of human interaction. That intimate interaction is often fraught with difficulties because of the emotional and physical difficulty of the work and low pay. Perhaps, if some tasks were taken over by a robot, more time would be freed up to actually communicate and interact with a lonely person.

VM

My quick reaction is this: Were I alive when some of the more general applications of telepresence become available, I would react in the same hesitant, behind-hand gradual way that I have to the onset of new computer capabilities in general over the last 50 or more years.

I didn't have a television of my own until I was 30 years, but I moved up to color only five years later. I didn't get a personal computer of my own until I was 46 years, long after I wrote about comsats and wafer fabs and photolithography and cell phones for *The New York Times* in the 1970s.

Your note touches on the sometimes unrecognized speed with which everyday circumstances change. The quarter-millennium fact of life is rapid generalizing of capabilities. Most of what keeps most of us alive are changes over the 10-generation period since Watt's invention in the 1760s of a separate-condenser steam engine achieving a fourfold saving in coal use from the Newcomen pump.

Once Watt and Boulton retired in 1800 with expiry of their extended patent, limitations on steam engines operating at more than one atmosphere pressure are lifted. We can think of Andrew Jackson migrating slowly, partly on horseback, from Tennessee to Washington to become President in 1829, and going back home eight years later on a train, or Lincoln, only a quarter-century later making the train journey both ways, alive going to Washington and dead coming back. The development of what can be called a "Networked house" follows in the cheap-steel rail era, which is also the oil era, the electric power era, the telephone daughter of the telegraph era, the era of screens on windows, indoor hot water and central heating, and mechanical refrigeration. A great many of these things involve a person acting indirectly, as in switching a flip or setting a dial.

An underlying principle, given my absent-minded professor mind, is to avoid attending to time-wasting distractions, starting with the game of bridge, but definitely including an excess of fantasy.

In general, I have a nineteenth-century attitude toward communication, that is, writing things out in advance clears thoughts and saves time in conversation. The corollary is that writing, properly conceived, can be conversational. I assume that this follows the Bismarckian principle that you should never enter a negotiation without knowing your aims clearly and never exceed these limits out of greed.

I have been using a hand-held cell phone device for about 20 years, largely as an emergency communication device, primarily with my wife and occasionally with doctors.

For the last six years, I have been using an iPhone 6S Plus. I expect to leap to one shy of the latest, that is to an iPhone 12 in 2022. These past 116 days, I used it to attest daily to the MIT Atlas system that I have no Covid symptoms. Since Labor Day, I have also physically brought in my self-test Covid ampoule, having photographed its serial number with my iPhone, and have each of 17 times received promptly the advice that I'm negative.

A succession of computers, starting with a DEC desktop on arrival at MIT in 1982, then a suitcase-size Compac, an IBM PC-XT, and, since 1990, one Apple desktop or laptop after another (now a 6-year-old MacBook Air) has allowed me far more fluent writing than was possible with lined paper in notebooks or pads or typing on a mechanical typewriter (as I did constantly in 20 years of active news reporting).

An irritation in recent years is the attempt to make driving a car or "tuning" a television set more automatic and impulsive. Too many electronic "smarts". But I do like having a small television camera that displays what's behind me when I put the lever in reverse.

MS

Managed well (safeguards on privacy, *not* interfering with others' lives, etc.), it could be a real boon in lots of areas. It could be not only a direct "net plus" in terms of safety, comfort, pleasure, etc. They could free up time to allow for individuals to pursue their own goals for education pleasure, leisure, etc.

Potentially a good evolution, but governance is a *real* issue, so as not to harm the "common good".

ER

Potential of that kind of world to live in is indeed very scary to me!!!!

It is scary because it leaves out human interactions and the quality of life that is enhanced by the dynamics of interchange with human beings. What has already happened because of the antisocial aspects of the impact of the virus is that there is an increase in mental health issues and for me, a sense of unease.

While I adore my great-grandchildren, I am very ambivalent about increasing the population and worry about the depersonalization of the world.

I doubt that I will live long enough to experience that world and am anxious about what it portends.

NH

I am essentially very, very wary. Although I can certainly see uses for such technology; in addition to those you mention, I think of assistance to disabled individuals. Its benefits will probably be technology that further isolates the haves from the have-nots of the world and could be an additional disruption to the loss of human jobs worldwide, leaving us with populations of aimless dissatisfied folks (as well as those who relish their free time and use it positively), and the governmental responsibility for seeing to the general welfare. But what worries me most is the potential for such "toys of power" to appeal most strongly to those who will put them to the least socially beneficial uses, including military and international moves. The stories we already hear about remote control drones and the moral and ethical damage they do while theoretically advancing "good clean warfare" are difficult to stomach. The next generation of gazillionaires may not be satisfied with orbiting the earth and shooting for the moon.

JG
We old folks grew up in another day when we enjoyed privacy that no longer exists. Telepresence unfortunately continues this loss of privacy. We were very fortunate (in many ways)! On We Go.

JF
Wow! This is very exciting. But I hope someone is thinking about security. I wouldn't want someone else messing with my robot.

BP
Whether we like it or not, the telepresence will be a reality. Like most scientific developments, it brings with it potential for good and for harm. How will it be regulated is a key question.

TB
If I want robots to perform boring tasks at my command, like deliver my coffee and NYT in bed every morning, cook my meals, clean my house, mow my lawn, rake my leaves, tend my roses, that's one thing.

If I use robots to perform some economic functions like plant or harvest crops, assemble goods, or perform intricate tasks like brain surgery, that's another.

But a robot that is an extension of my warped personality in other dimensions is a puzzle. If I say or think, "I hate Trump. Someone should kill him!", would my avatar do it? (Imagine the avatar traffic jam at Mar El Lago.) Would it do other harmful acts imagined by me? Am I criminally or civilly liable for acts of my avatar?

If I get angry and act out verbally or physically, does my avatar do so as well? If I suffer from depression or other mental issues, the same for my avatar? Memory loss as well? What happens if I forget my avatar, does my avatar forget me and become an independent being or just sit in a corner or turn off?

Finally, why would I want an avatar? Is not one of me more than enough?

Happy new year to all and your avatars!

AB
Scary. What's to prevent it from becoming the master in your relationship?

PK
I would want to minimize telepresence in my life out of concern for the possibility of weakening my own muscles, coordination, resolve, and even feelings. If I ever become entirely immobile, I might find the assistance of a robot in fetching things for me helpful, but I want to be "at the table" with my entire self—for meetings, socializing, and physically doing what I can.

PS
My reactions are all over the place. I was once a social worker dealing with families with severe problems, including disturbed and abused children, so that was the "job" that came to mind. Such a robot would enable close supervision of volatile situations, but this feels like a gross violation of privacy. On the other hand, it could also enable timely consultation and coaching to parents at times of crisis. The results, as

always, would depend on the skill and judgment of the person doing the job. It feels fundamentally creepy. It would also be very demanding because it could require 24/7 availability and would bring clinicians into the worst of situations. It might in itself help parents by giving them a sense of continuous support. They might also develop much resentment because of intrusive control. It might be like having a live-in grandmother, with all the good and bad aspects.

On an everyday level, such a robot could provide a way for all grandparents to provide support at a distance. Again, its usefulness and acceptability would depend on the dynamics of the whole family situation.

As I think about it more, I wonder what this would really add to the current use of cell phones. My work in this area ended almost 40 years ago, so I'm way out of date. On another, completely different note, it would be a wonderful way to travel virtually when you're no longer able to travel physically.

SL

I first wondered if your avatar could think, and if it could think would it do so independent of you. I worry. Perhaps it could not think initially, but will learn as it needs to make decisions even when it is not being controlled. Can it reproduce? Make itself with your instructions and then if it learns, it may reproduce without your instructions. Would that be like synthesizing humans from individual cells? I think that your book might be dangerous if it falls into the wrong hands. Reconsider!

PM

I see many new advantages for people who are disabled, immobile, desiring to paint or knit or garden, curious, wanting to cook and stay independent, and investigating various laboratory sciences and medical research. All wonderful opportunities. On the other hand, given man's ability to disturb and destroy other peoples and societies, I am frightened by the terrible events the robots can facilitate and perform themselves. Can regulation be effective in controlling these dangerous powers?

MC

My reaction is one of total approval! This would be a fabulous advance in civilization and would remove much of the backbreaking drudgery needed to support our lives. I am particularly interested in the use of such robots in manufacturing since they would enable low-cost production locally, which would be beneficial to both the economy and the ecology. As in any new technology, there would be concerns about physical safety and hacking, so that safeguards would have to be built in. There would also be societal implications to the widespread use of such robots, so that efforts would be needed to educate both the users and the public, while many workers would have to be trained for new employment opportunities.

LD

Reproduction has been essential to life forever. Surely such telepresence technology would be a part for components of it. From the convenience of partners separated by oceans or interplanetary space to evil uses such as rape of enemies. Thankfully, I won't be around for any of it.

CD
Currently, I multitask unsuccessfully, so an avatar would not be helpful. As I understand it, I, alone, would be monitoring it constantly and it could even duplicate some human feelings and emotions that I might have. However, my avatar could never satisfy my own desire to touch or kiss a loved one, or sing a song. So, since I cannot depend on my avatar at this stage of my life to sustain connections to loved ones, it is of no use to me.

Of course, I understand its potential value in science and medicine. I can also envision a world of billions of people and their robots during a catastrophic occurrence due to climate change or something else. In that scenario, the avatars would be of no value.

KS
Fascinating, empowering.

LL
I feel a mixture of dread and sadness. For science and the military, these advances can have benefits that far outweigh the downsides. For our everyday life, it does the opposite: making us more and more disconnected from reality and from each other at a time when we desperately need the reverse. It will clearly lead to the end of civilization as we know it, and I am sorry for our grandchildren.

AW
My immediate reaction is that this technological development will in the short-run (25 years??) increase the distancing of interpersonal relationships as it has for so many of our grandchildren. I fear that it will also increase the divide between the rich and the poor—the educated and uneducated.

Maybe in the more distant future, it will increase our freedom to contribute to arts and environmental needs, in that it frees up our time for more creative pursuits and helps improve our world for the good of all.

HW
Sounds like the bravest of new worlds. I hope that its development is slow enough to allow society time to absorb the changes it will bring. Lastly, I worry about the evil that will accompany the adoption and use of telepresence.

Appendix 2: Bibliography

The following pages offer an extensive bibliography, with emphasis on more recent publications that are relevant to telepresence and virtual reality. The listing is unavoidably biased toward research in virtual reality because most recent telerobotics research is on functioning stand-alone robots, especially making use of artificial intelligence, a topic that is not emphasized in this book.

I have clustered the listings into the following salient categories and subcategories that I hope will make this more useful to the reader/researcher:

1. **Implementing telepresence** (Robot configurations; Using avatars in VR; Problems of control, delay; Software requirements; Measuring sense of telepresence)
2. **Human operator sensing and information display** (Vision and visual occlusion: Depth perception; HMD design, Touch, Data gloves; Haptics; Kinesthetics and gesture recognition, Position tracking; Master control devices)
3. **Teleoperation** (Telesurgery; Teleoperated assembly; Flight simulation; Driving simulation)
4. **Group interaction** (2-person telecollaboration; Multi-person teleconferencing; Learning; Social behavior; Theater; Gaming)
5. **Assistive function** (Physically disabled; Social robotics)
6. **Commercial activities in VR** (Shopping; Marketing; Tourism)
7. **Legal, ethical, regulatory, and philosophical issues** (Legal; Philosophical)

IMPLEMENTING TELEPRESENCE

Robot Configurations

Adamides G, Christou G, Katsanos C, Xenos M and Hadzilacos T: Usability guidelines for the design of robot teleoperation: A taxonomy, *IEEE Transactions on Human-Machine Systems* 45(2): 256–262, 2015.

Fritsche L, Unverzag F, Peters J and Calandra R: First-person teleoperation of a humanoid robot, in *Proceedings of the 2015 IEEE-RAS 15th International Conference Humanoid Robots*, pp. 997–1002, 2015.

Miller N, Jenkins O, Kallmann M and Mataric M: Motion capture from inertial sensing for untethered humanoid teleoperation, in *Proceedings of the 4th IEEE/RAS International Conference Humanoid Robots* 2, pp. 547–565, 2004.

Sheridan T: Human-robot interaction: Status and challenges, *Human Factors* 58(4): 525–532, 2016.

Sheridan T: *Telerobotics, Automation and Human Supervisory Control*, Cambridge, MA, MIT Press, 1992.

Using Avatars in VR

Argelaguet F, Hoyet L, Trico M and Lecuyer A: The role of interaction in virtual embodiment: Effects of the virtual hand representation, in *2016 IEEE Virtual Reality (VR)*, pp. 3–10, 2016.

Blanz I and Vetter T: A morphable model for the synthesis of 3d faces, in *1999 Proceedings of the 26th Annual Conference on Computer Graphics and Interactive Techniques*, Boston, MA, ACM Press/Addison-Wesley Publishing Co., pp. 187–194.

Burgos-Artizzu X, Fleureau J, Dumas O, et al: Real-time expression-sensitive HMD face reconstruction, in *SIGGRAPH Asia Technical Briefs, ACM* 9, 2015.

Cai Q, Gallup D, Zhang C, and Zhang Z: 3d deformable face tracking with a commodity depth camera, in *Computer Vision–ECCV*, Springer, pp. 229–242, 2010.

Cao C, Cao X, Wei Y, Wen F, and Sun J: Face alignment by explicit shape regression, *International Journal of Computer Vision* 107(2): 177–190, 2014.

Greenwald S, Wang Z, Funk M and Maes P: Investigating social presence and communication with embodied avatars in room-scale virtual reality, in *International Conference on Immersive Learning*, pp. 75–90, 2017.

Grubert J, Witzani L, Ofek E, Pahud M, et al: Effects of hand representations for typing in virtual reality, in *2018 IEEE Conference on Virtual Reality and 3D User Interfaces (VR)*, pp. 151–158, 2018.

Heidicker P, Langbehn E and Steinicke F: Influence of avatar appearance on presence in social VR, in *2017 IEEE Symposium on 3D User Interfaces (3DUI)*, pp. 233–234, 2017.

Hoyet L, Argelaguet F, Nicole C and Lécuyer A: Wow! I have six fingers!: Would you accept structural changes of your hand in VR? *Frontiers in Robotics and AI* 3: 27, 2016.

Huang W, Alem L, Tecchia F and Duh B: Augmented 3d hands: A gesture-based mixed reality system for distributed collaboration, *Journal on Multimodal User Interfaces* 12(2): 77–89, 2018.

Jo D, Kim K and Kim G: Effects of avatar and background representation forms to co-presence in mixed reality (mr) tele-conference systems, in *SIGGRAPH ASIA Virtual Reality meets Physical Reality: Modelling and Simulating Virtual Humans and Environments*, pp. 12, 2016.

Jo D, Kim K and Kim G: Effects of avatar and background types on users' co-presence and trust for mixed reality-based teleconference systems, in *Proceedings the 30th Conference on Computer Animation and Social Agents*, pp. 27–36, 2017.

Jung S, Wisniewski P and Hughes C: In limbo: The effect of gradual visual transition between real and virtual on virtual body ownership illusion and presence, in *Proceedings of the 25th IEEE Conference on Virtual Reality and 3D User Interfaces IEEE VR*, 2018.

Latoschik M, Roth D, Gall D, Achenbach J, et al: The effect of avatar realism in immersive social virtual realities, in *Proceedings of the 23rd ACM Symposium on Virtual Reality Software and Technology*, pp. 39, 2017.

Lin L and Jörg S: Need a hand? How appearance affects the virtual hand illusion, in *Proceedings of the ACM Symposium on Applied Perception*, pp. 69–76, 2016.

Lin L, Normovle A, Adkins A, Sun Y, et al: The effect of hand size and interaction modality on the virtual hand illusion, in *2019 IEEE Conference on Virtual Reality and 3D User Interfaces (VR)*, pp. 510–518, 2019.

Lugrin J, Ertl M, Krop P, Klupfel R, et al: Any body there? Avatar visibility effects in a virtual reality game, in *2018 IEEE Conference on Virtual Reality and 3D User Interfaces (VR)*, pp. 17–24, 2018.

Lugrin J, Latt J and Latoschik M: Avatar anthropomorphism and illusion of body ownership in VR, in *Virtual Reality (VR) 2015 IEEE*, pp. 229–230, 2015.

Lugrin J, Wiedemann M, Bieberstein D and Latoschik M: Influence of avatar realism on stressful situation in VR, in *2015 IEEE Virtual Reality (VR)*, pp. 227–228, 2015.

Medeiros D, dos Anjos R, Mendes D, et al: Keep my head on my shoulders! Why third-person is bad for navigation in VR, in *Proceedings of the 24th ACM Symposium on Virtual Reality Software and Technology VRST '18*, pp. 16:1–16:10, 2018.

Nagendran A, Steed A, Kelly B and Pan Y: Symmetric telepresence using robotic humanoid surrogates, *Computer Animation and Virtual Worlds* 26(3–4): 271–280, 2015.

Noh S, Yeo H and Woo W: An HMD-based mixed reality system for avatar-mediated remote collaboration with bare-hand interaction, in *ICAT-EGVE2015—International Conference on Artificial Reality and Telexistence and Eurographics Symposium on Virtual Environments*, 2015.

Oda O, Elvezio C, Sukan M, Feiner S and Tversky B: Virtual replicas for remote assistance in virtual and augmented reality, in *Proceedings of the 28th Annual ACM Symposium on User Interface Software & Technology*, pp. 405–415, 2015.

Ogawa N, Narumi T and Hirose M: Virtual hand realism affects object size perception in body-based scaling, *2019 IEEE Conference on Virtual Reality and 3D User Interfaces (VR)*, pp. 519–528, 2019.

Pan Y and Steed A: The impact of self-avatars on trust and collaboration in shared virtual environments, *PloS one* 12(12): e0189078, 2017.

Reich R, Sahami M and Weinstein J: *System Error: Where Big Tech Went Wrong and How We Can Reboot*, Harper-Collins, 2021.

Reinhard E, et al: Color transfer between images, *IEEE Computer Graphics and Applications* 5: 34–41, 2001.

Rizzo A, Neumann U, Enciso R, Fidaleo D and Noh J: Performance-driven facial animation: Basic research on human judgments of emotional state in facial avatars, *Cyberpsychology & behavior* 4(4): 47, 2001.

Rosenberg L: Virtual fixtures: Perceptual tools for telerobotic manipulation, in *Proceedings of IEEE Virtual Reality Annual International Symposium*, pp. 76–82, 1993.

Roth D, Waldow K, Latoschik M, Fuhrmann A and Bente G: Socially immersive avatar-based communication, in *Virtual Reality (VR) 2017 IEEE*, pp. 259–260, 2017.

Saraiji M, Sasaki T, Matsumura R, Minamizawa K and Inami M: Fusion: Full body surrogacy for collaborative communication, in *ACM SIGGRAPH 2018 Emerging Technologies*, pp. 1–2, 2018.

Schwind V, Knierim P, Chuang L and Henze N: Where's pinky? The effects of a reduced number of fingers in virtual reality, in *Proceedings of the Annual Symposium on Computer-Human Interaction in Play*, pp. 507–515, 2017.

Schwind V, Knierim P, Tasci C, Franczak P, et al: These are not my hands! Effect of gender on the perception of avatar hands in virtual reality, in *Proceedings of the 2017 CHI Conference on Human Factors in Computing Systems*, pp. 1577–1582, 2017.

Steptoe W, Steed A, Rovira A, and Rae J: Lie tracking: Social presence, truth and deception in avatar-mediated telecommunication, in *Proceedings of the SIGCHI Conference on Human Factors in Computing Systems, ACM*, pp. 1039–1048, 2010.

Waltemate T, Gall D, Roth D, Botsch M, Latoschik M: How foot tracking matters: The impact of an animated self-avatar on interaction, embodiment and presence in shared virtual environments, *IEEE Transactions on Visualization and Computer Graphics* 24(4): 1643–1652, 2018.

Wauck H, Lucas G, Shapiro A, Feng A, et al: Analyzing the effect of avatar self-similarity on men and women in a search and rescue game, in *Proceedings of the 2018 CHI Conference on Human Factors in Computing Systems*, pp. 485, 2018.

Weng Y, Zhou S, Tong Y and Zhou K: Facewarehouse: A 3D facial expression database for visual computing, *IEEE Transactions on Visualization and Computer Graphics* 20(3): 413–425, 2014.

Xiao J, Baker S, Matthews I and Kanade T: Real-time combined 2d+ 3d active appearance models, *CVPR* 2: 535–542, 2004.

Yoon B, Kim H, Lee G, Billinqhurst M and Woo W: The effect of avatar appearance on social presence in an augmented reality remote collaboration, in *2019 IEEE Conference on Virtual Reality and 3D User Interfaces (VR)*, pp. 547–556, 2019.

Yoon B, Kim H and Woo W: Evaluating remote virtual hands models on social presence in hand-based 3D remote collaboration, in *19th IEEE International Symposium on Mixed and Augmented Reality (ISMAR)*, pp. 520–532, 2020.

Yu K, Gorbachev G and Roth D: Avatars for teleconsultation: Effects of avatar embodiment techniques on user perception in 3D asymmetric telepresence, in *IEEE Transactions on Visualization and Computer Graphics* 27(11): 4129–4139, November 2021.

Zhao Y: Mask-off: Synthesizing face images in the presence of head-mounted displays, in *26th IEEE Conference on Virtual Reality and 3D User Interfaces (VR)*, pp. 267–276, 2015, 2019.

Problems with Control, Delay, and Cybersickness

Bejczy A, Kim W and Venema S: The phantom robot: Predictive displays for teleoperation with time delay, in *Proceedings of the IEEE International Conference on Robotics and Automation*, pp. 546–551, 1990.

Chang S, Kim I, Borm J, Lee C and Park J: KIST teleoperation system for humanoid robot, in *Proceedings of the IEEE/RSJ International Conference on Intelligent Robots and Systems* 2, pp. 1198–1203, 1999.

Chen S, Quicktime VR: An image-based approach to virtual environment navigation, in *Proceedings of the 22nd Annual Conference on Computer Graphics and Interactive Techniques*, pp. 29–38, 1995.

Fernando C, et al: Design of TELESAR V for transferring bodily consciousness in telexistence, in *Proceedings of the 2012 IEEE/RSJ International Conference on Intelligent Robots and Systems*, pp. 5112–5118, 2012.

Goldberg K, Mascha M, Gentner S, Rothenberg N, et al: Desktop teleoperation via the world wide web, in *Proceedings of the IEEE International Conference on Robotics and Automation* 1, pp. 546–551, 1995.

Liu R, Zhuang C, Yang R and Ma L: Effect of economically friendly acustimulation approach against cybersickness in video-watching tasks using consumer virtual reality devices, September 2019, *Applied Ergonomics*, 82, 2019.

Sayers C, Paul R, Whitcomb L and Yoerger D: Teleprogramming for subsea teleoperation using acoustic communication, *IEEE Journal of Oceanic Engineering* 23(1): 60–71, 1998.

Weech S, Kenny S and Barnett-Cowan M: Presence and cybersickness in virtual reality are negatively related: A review, *Frontiers in Psychology* 10, pp. 158, 2019.

Geometry and Software Requirements to Achieve Telepresence

Brilakis I, Fathi H and Rashidi A: Progressive 3d reconstruction of infrastructure with videogrammetry, *Automation in Construction* 20(7): 884–895, 2011.

Cernigliaro G, Martos M, Montagud M, et al: PC-MCU: Point cloud multipoint control unit for multi-user holoconferencing systems, in *NOSSDAV '20: Proceedings of the 30th ACM Workshop on Network and Operating Systems Support for Digital Audio and Video*, pp. 47–53, 2020.

Chen J, Bautembach D and Izadi S: Scalable real-time volumetric surface reconstruction, *ACM Transactions on Graphics* 32(4): 1–16, 2013.

Correll K, Barendt N and Branicky M: Design considerations for software only implementations of the IEEE 1588 precision time protocol, in *Proceedings of the Conference IEEE* 1588, pp. 11–15, 2005.

Debevec P: Rendering synthetic objects into real scenes: Bridging traditional and image-based graphics with global illumination and high dynamic range photography, in *Proceedings of the 25th Annual Conference on Computer Graphics and Interactive Techniques*, pp. 189–198, 1998.

Dou M: Fusion 4D: Real-time performance capture of challenging scenes, *ACM Transactions on Graphics* 35(4): 1–13, 2016.

Fuchs H, State A and Bazin J: Immersive 3D telepresence, *Computer* 47(7): 46–52, 2014.

Gao L, Bai H, Lindeman R and Billinghurst M: Static local environment capturing and sharing for mr remote collaboration, in *SIGGRAPH Asia 2017 Mobile Graphics & Interactive Applications*, pp. 17, 2017.

Golodetz S, Cavallari T, Lord N, Prisacariu A, et al: Collaborative large-scale dense 3D reconstruction with online inter-agent pose optimisation, *IEEE Transactions on Visualization and Computer Graphics* 24(11): 2895–2905, November 2018.

Hvass J, Larsen O, Vendelbo K, Nilsson N, et al: The effect of geometric realism on presence in a virtual reality game, in *Virtual Reality (VR) 2017 IEEE*, pp. 339–340, 2017.

Iorns T and Rhee T: Real-time image based lighting for 360-degree panoramic video, in *Pacific-Rim Symposium on Image and Video Technology*, pp. 139–151, 2015.

Lensing P and Broll W: Instant indirect illumination for dynamic mixed reality scenes, in *Mixed and Augmented Reality (ISMAR) 2012 IEEE International Symposium on*, pp. 109–118, 2012.

Measuring Sense of Telepresence

Biocca F, Harms C and Gregg J: The networked minds measure of social presence: Pilot test of the factor structure and concurrent validity, in *4th Annual International Workshop on Presence*, pp. 1–9, 2001.

Biocca F, Harms C and Burgoon J: Toward a more robust theory and measure of social presence: Review and suggested criteria, *Presence: Teleoperators and Virtual Environments* 12(5): 456–480, October 2003.

Bouchard S, Dumoulin S, Talbot J, Ledoux A, et al: Manipulating subjective realism and its impact on presence: Preliminary results on feasibility and neuroanatomical correlates, *Interacting with Computers* 24(4): 227–236, 2012.

Bulu S: Place presence social presence co-presence and satisfaction in virtual worlds, *Computers & Education* 58(1): 154–161, 2012.

Buttussi F and Chittaro L: Effects of different types of virtual reality display on presence and learning in a safety training scenario, in *IEEE Transactions on Visualization and Computer Graphics* 24(2): 1063–1076, 2017.

Draper J, Kaber D and Usher J: Telepresence, *Human Factors* 40(3): 354–375, 1998.

Fribourg R, Argelaguet F, Hoyet L and Lécuyer A: Studying the sense of embodiment in VR shared experiences, in *VR 2018 – 25th IEEE Conference on Virtual Reality and 3D User Interfaces*, pp. 1–8, March 2018.

Harms C and Biocca F: Internal consistency and reliability of the networked minds measure of social presence, in *Seventh Annual International Workshop: Presence 2004*, pp. 246–251, 2004.

Heeter C: Being there: The subjective experience of presence, *Presence: Teleoperators & Virtual Environments* 1(2): 262–271, 1992.

Ijsselsteijn W, de Ridder H, Freeman J and Avons S: Presence: Concept determinants and measurement, *Human Vision and Electronic Imaging V* 3959: 520–530, 2000.

Nowak K and Biocca F: The effect of the agency and anthropomorphism on users' sense of telepresence copresence and social presence in virtual environments, *Presence: Teleoperators & Virtual Environments* 12(5): 481–494, 2003.

Orus C, Ibanez-Sanchez S and Flavian C: Enhancing the customer experience with virtual and augmented reality: The impact of content and device type, *International Journal of Hospitality Management* 98(September) 2021.

Schubert T, Friedmann F and Regenbrecht H: The experience of presence: Factor analytic insights, *Presence: Teleoperators & Virtual Environments* 10(3): 266–281, 2001.

Sheridan T: Musings on telepresence and virtual presence, *Presence: Teleoperators & Virtual Environments* 1(1): 120–126, 1992.

Steptoe W, Steed A, Rovira A and Rae J: Lie tracking: Social presence, truth and deception in avatar-mediated telecommunication, in *Proceedings of the SIGCHI Conference on Human Factors in Computing Systems, ACM*, pp. 1039–1048, 2010.

Steuer J: Defining virtual reality: Dimensions determining telepresence, *Journal of Communication* 42(4): 73–93, 1992.

Vorderer P, Wirth W, Gouveia F, Biocca F, et al: Mec spatial presence questionnaire, *Retrieved Sept* 18: 2015, 2004.

Witmer B and Singer M: Measuring presence in virtual environments: A presence questionnaire, *Presence* 7(3): 225–240, 1998.

Zahorik P and Jenison R: Presence as being-in-the-world, *Presence* 7(1): 78–89, 1998.

Zhao S: Toward a taxonomy of copresence, *Presence: Teleoperators & Virtual Environments* 12(5): 445–455, 2003.

HUMAN OPERATOR SENSING AND INFORMATION DISPLAY

Visual Perception and Occlusion

Adcock M, Anderson S and Thomas B: Remote fusion: Real time depth camera fusion for remote collaboration on physical tasks, in *Proceedings of the 12th ACM SIGGRAPH International Conference on Virtual-Reality Continuum and Its Applications in Industry*, pp. 235–242, 2013.

Adelson E, Anderson C, Bergen J, Burt P and Ogden J: Pyramid methods in image processing, *RCA Engineer* 29(6): 33–41, 1984.

Besl P and Mckay N: Method for registration of 3-d shapes, in *Robotics-DL Tentative, International Society for Optics and Photonics*, pp. 586–606, 1992.

Daugman J: How iris recognition works, *IEEE Transactions on Circuits and Systems for Video Technology* 14(1): 21–30, 2004.

Gruber L, Ventura J and Schmalstieg D: Image-space illumination for augmented reality in dynamic environments, in *Virtual Reality (VR) 2015 IEEE*, pp. 127–134, 2015.

Handa A, Whelan T, McDonald J, and Davison A: A benchmark for RGB-D visual odometry, 3D reconstruction and SLAM, in *Proceedings of the International Conference on Robotics and Automation*, pp. 1524–1531, 2014.

Henry P, Fox D, Bhowmik A and Mongia R: Patch volumes: Segmentation-based consistent mapping with RGB-D cameras, in *The International Conference on 3D Vision*, 29 June, 2013.

Huang G, Ramesh M, Berg T and Learnedmiller E: Labeled faces in the wild: A database for studying face recognition in unconstrained environments, Tech. Rep. 07-49, University of Massachusetts, Amherst, October, 2007.

Izadi S, et al: Kinect fusion; real-time 3D reconstruction and interaction using a moving depth camera, in *Proceedings of the ACM Symposium on User Interface Software and Technology*, pp. 559–568, 2011.

Jones B, et al: Room alive, magical experiences enabled by scalable, adaptive projector-camera units, in *Proceedings of the Annual Symposium on User Interface Software and Technology*, pp. 637–644, 2014.

Kahler O, Prisacariu V and Murray D: Real-time large-scale dense 3D reconstruction with loop closure, in *European Conference on Computer Vision*, pp. 500–516, 2016.

Kahler O, Prisacariu J, Valentin J and Murray D: Hierarchical voxel block hashing for efficient integration of depth images, in *IEEE Robotics and Automation Letters*, pp. 192–197, 2016.

Kazemi V and Sullivan J: One millisecond face alignment with an ensemble of regression trees, in *Proceedings of the IEEE Conference on Computer Vision and Pattern Recognition*, pp. 1867–1874, 2014.

Keller M, Lefloch D, Lambers M, Izadi S, et al: Real-time 3D reconstruction in dynamic scenes using point-based fusion, in *Proceedings of Joint 3DIM/3DPVT Conference*, pp. 8, 2013.

Kim J and Interrante V: Dwarf or giant: the influence of interpupillary distance and eye height on size perception in virtual environments, in *International Conference on Artificial Reality and Telexistence Eurographics Symposium on Virtual Environments*, R Lindeman, G Bruder, and D Iwai (Editors), 2017.

Komiyama R, Miyaki T and Rekimoto J: JackIn space: Designing a seamless transition between first and third person view for effective telepresence collaborations, in *Proceedings of the 8th Augmented Human International Conference, ACM* 14: 1–9, March 2017.

Levin A, Lischinski D and Weiss Y: Colorization using optimization, *ACM Transactions on Graphics (TOG)* 23: 689–694, 2004.

Lorensen W and Cline H: Marching cubes; a high resolution 3D surface construction algorithm, in *Proceedings of the 14th Annual Conference on Computer Graphics and Interactive Techniques*, pp. 163–169, 1987.

MacQuarrie A and Steed A: Cinematic virtual reality: Evaluating the effect of display type on the viewing experience for panoramic video, in *IEEE Virtual Reality (VR)*, pp. 45–54, 2017.

Maier R, Schaller R and Cremers D: Efficient online surface correction for real-time large-scale 3D reconstruction, in *British Machine Vision Conference (BMVC)*, September 2017.

Maimone A, Bidwell J, Peng K and Fuchs H: Enhanced personal autostereoscopic telepresence system using commodity depth cameras, *Computers & Graphics* 36(7):791–807, 2012.

Maimone A and Fuchs H: Real-time volumetric 3D capture of roomsized scenes for telepresence, in *Proceedings of the 3DTV-Conference*, October 2012.

Medeiros D, Sousa M, Mendes D, Raposo A and Jorge J: Perceiving depth: optical versus video see-through, in *Proceedings of the 22nd ACM Conference on Virtual Reality Software and Technology*, pp. 237–240, 2016.

Molyneaux D, Izadi S, Kim D, Hilliges O, et al: Interactive environment-aware handheld projectors for pervasive computing spaces, in *Proceedings of the International Conference on Pervasive Computing*, pp. 197–215, 2012.

Mossel A and Kroter M: Streaming and exploration of dynamically changing dense 3D reconstructions in immersive virtual reality, in *Proceedings of IEEE International Symposium on Mixed and Augmented Reality*, pp. 43–48, 2016.

Newcombe R, Fox D, and Seitz S: Dynamic fusion: Reconstruction and tracking of non-rigid scenes in real-time, in *IEEE Conference on Computer Vision and Pattern Recognition*, pp. 343–352, 2015.

Ng A, Chan L and Lau H: Depth perception in virtual environment: The effects of immersive system and freedom of movement, in *International Conference on Virtual, Augmented and Mixed Reality*, New York: Springer, 173–183, 2016.

Nießner M, Zollhofer M, Izadi S and Stamminger M: Real-time 3D reconstruction at scale using voxel hashing, *ACM Transactions on Graphics* 32(6): 1–11, 2013.

Olszewski K, Lim J, Saito S and Li H: High fidelity facial and speech animation for VR HMDs, in *ACM Transactions on Graphics (Proceedings SIGGRAPH Asia)* 35(6) December, 2016.

Pardo P, Suero M and Pérez A: Correlation between perception of color shadows and surface textures and the realism of a scene in virtual reality, *JOSA A* 35(4): B130–B135, 2018.

Parke, F and Waters K: *Computer Facial Animation*, A. K. Peters, 1996.

Pejsa T, Kantor J, Benko H, Ofek E and Wilson A: Room2room enabling life-size telepresence in a projected augmented reality environment, in *Proceedings of the 19th ACM Conference on Computer-Supported Cooperative Work & Social Computing CSCW '16*, pp. 1716–1725, 2016.

Pintaric T, Neumann U and Rizzo A: Immersive panoramic video, in *Proceedings of the 8th ACM International Conference on Multimedia* 493, 2000.

Reichl F, Weiss J and Westermann R: Memory-efficient interactive online reconstruction from depth image streams, *Computer Graphics Forum* 35(8): 108–119, 2016.

Reinhard E, Ashikhmin M, Gooch B and Shirley P: Color transfer between images, *IEEE Computer Graphics and Applications* 5: 34–41, 2001.

Romera-Paredes B, Zhang C and Zhang Z: Facial expression tracking from head-mounted, partially observing cameras, in *Multimedia and Expo (ICME), IEEE International Conference on, IEEE*, pp. 1–6, 2014.

Roth H and Vona M: Moving volume kinectfusion, in *Proceedings of the British Machine Vision Conference*, pp. 112.1–112.11, 2012.

Slater M, Usoh M and Steed A: Depth of presence in virtual environments, *Presence: Teleoperators & Virtual Environments* 3(2): 130–144, 1994.

Steed A, Steptoe W, Oyekoya W, Pece F, et al: Beaming, An asymmetric telepresence system, *IEEE Computer Graphics and Applications* 32(6): 10–17, November 2012.

Stotko P, Krumpen S, Weinmann M and Klein R: SLAMCast: Large-scale, real-time 3D reconstruction and streaming for immersive multi-client live telepresence. *IEEE Transactions on Visualization and Computer Graphics* 25(5): 2102–2112, 2019.

Tang A, Fakourfar O, Neustaedter C and Bateman S: *Collaboration in 360 Videochat: Challenges and Opportunities*, Proc, Conference on Designing Interactive Systems, pp. 1327–1339, June 2017.

Thanyadit S, Punpongsanon P and Pong T: Observar, Visualization system for observing virtual reality users using augmented reality, in *2019 IEEE International Symposium on Mixed and Augmented Reality (ISMAR)*, pp. 258–268, 2019.

Thies J, Zollhofer M, Niessner M, Valgaerts L, Stamminger M and Theobalt C: Real-time expression transfer for facial reenactment, *ACM Transactions on Graphics (TOG)* 34(6): 1–14, 2015.

Tran H, Ngoc N, Pham C, Jung Y and Thang T: A subjective study on QoE of 360 video for VR communication, *IEEE 19th International Workshop on Multimedia Signal Processing (MMSP)*, pp. 1–6, 2017.

Vasudevan A, Kurillo G, Lobaton E, et al: High-quality visualization for geographically distributed 3-D teleimmersive applications, *IEEE Transactions on Multimedia* 13(3): 573–584, 2011.

Whelan T, Johannsson H, Kaess M, Leonard J and McDonald J: Robust real-time visual odometry for dense RGB-D mapping, in *IEEE International Conference on Robotics and Automation*, pp. 5724–5731, 2013.

Whelan T, Kaess M, Johannsson H, Fallon M, Leonard J and McDonald J: Real-time large-scale dense RGB-D SLAM with volumetric fusion, *The International Journal of Robotics Research* 34(4–5): 598–626, 2015.

Xiao J, Baker S, Matthews I and Kanade T: Real-time combined 2d+ 3d active appearance models, *CVPR* 2: 535–542, 2004.

Yee Y: *Spatiotemporal Sensitivity and Visual Attention for Efficient Rendering of Dynamic Environments*. Master's thesis, Cornell University, 2000.

Zhang W, Shan S, Chen X and Gao W: Local gabor binary patterns based on Kullback–Leibler divergence for partially occluded face recognition, *Signal Processing Letters, IEEE* 14(11): 875–878, 2007.

Zhao Y, Huang X, Gao J, Tokuta A, Zhang C and Yang R: Video face beautification, in *2014, ICME*, pp. 1–6, 2014.

Mixed Reality, Head-Mounted Display

Anjos R, Sousa M, Mendes D, Medeiros D, et al: Adventures in hologram space: Exploring the design space of eye-to-eye volumetric telepresence, in *25th ACM Symposium on Virtual Reality Software and Technology VRST 19*, Association for Computing Machinery, 2019.

De Pace F, Manuri F, Sanna A and Zappia D: *Frontiers of Robotics AI*. July 2019.

Fairchild A, Campion S, Garcıa A, Wolff R, et al: A mixed reality telepresence system for collaborative space operation, *IEEE Transactions on Circuits and Systems for Video Technology* 27(4): 814–827, 2016.

Fonseca D and Kraus M: A comparison of head-mounted and hand-held displays for 360 videos with focus on attitude and behavior change, *Proceedings of the 20th International Academic Mindtrek Conference*, pp. 287–296, 2016.

Fraustino J, Lee Y, Lee S and Ahn H: Effects of 360 video on attitudes toward disaster communication: Mediating and moderating roles of spatial presence and prior disaster media involvement, *Public Relations Review* 44(3): 331–341, 2018.

Hartley R and Zisserman A: Multiple view geometry in computer vision, 2003. Cambridge University Press. Microsoft hololens. http://www.microsoft.com/microsoft-hololens/en-us. Accessed: 2015-11-05.

Kartch D: Efficient rendering and compression for full parallax computer-generated holographic stereograms, 2000. PhD thesis, Cornell University.

Kim S, Lee G, Huang W, Kim H, et al: Evaluating the combination of visual communication cues for HMD-based mixed reality remote collaboration, in *Proceedings of the 2019 CHI Conference on Human Factors in Computing Systems*, pp. 173, 2019.

Knierim P, Schwind V, Feit A, Nieuwenhuizen F and Henze N: Physical keyboards in virtual reality: Analysis of typing performance and effects of avatar hands, in *Proceedings of the 2018 CHI Conference on Human Factors in Computing Systems*, pp. 345, 2018.

Kruijff E, Swan J and Feiner S: Perceptual issues in augmented reality revisited, in *2010 IEEE International Symposium on Mixed and Augmented Reality*, pp. 3–12, 2010.

Lee C, Rincon G, Meyer G, Höllerer T and Bowman D: The effects of visual realism on search tasks in mixed reality simulation, *IEEE Transactions on Visualization and Computer Graphics* 19(4): 547–556, 2013.

Livingston M, et al: Basic perception in head-worn augmented reality displays, Chapter 1, in *Human Factors in Augmented Reality Environments*, New York: Springer, 2013.

Lukosch S, Lukosch H, Datcu D and Cidota M: Providing information on the spot: Using augmented reality for situational awareness in the security domain, *Computer Supported Cooperative Work (CSCW)* 24(6): 613–664, December 2015.

Norman M, Lee G, Smith R and Billingurst M: The impact of remote user's role in a mixed reality mixed presence system, in *The 17th International Conference on Virtual-Reality Continuum and its Applications in Industry*, pp. 1–9, 2019.

Orts-Escolano S, Rhemann C, Fanello S, Chang W, et al: Holoportation: Virtual 3D teleportation in real-time, in *Proceedings of the 29th Annual Symposium on User Interface Software and Technology*, pp. 741–754, 2016.

Pagani A and Stricker D: Structure from motion using full spherical panoramic cameras, in *2011 IEEE International Conference on Computer Vision Workshops (ICCV Workshops)*, pp. 375–382, 2011.

Pan Y, Sinclair D and Mitchell K: Empowerment and embodiment for collaborative mixed reality systems, *Computer Animation and Virtual Worlds* 29(3–4): e1838, 2018.

Pejsa T, Kantor J, Benko H, Ofek E and Wilson A: Room2room: Enabling life-size telepresence in a projected augmented reality environment, in *Proceedings of the 19th ACM Conference on Computer-Supported Cooperative Work & Social Computing*, pp. 1716–1725, 2016.

Piumsomboon T, Day A, Ens B, Lee G and Billinghurst M: Exploring enhancements for remote mixed reality collaboration, in *SIGGRAPH Asia 2017 Mobile Graphics & Interactive Applications*, pp. 16, 2017.

Piumsomboon T, Dey A, Ens B, Lee G and Billinghurst M: The effects of sharing awareness cues in collaborative mixed reality, *Frontiers in Robotics and AI* 6: 5, 2019.

Piumsomboon T, Lee G, Hart J, Ens B, et al: Mini-me: An adaptive avatar for mixed reality remote collaboration, in *Proceedings of the 2018 CHI Conference on Human Factors in Computing Systems*, pp. 46:1–46:13, 2018.

Piumsomboon T, Lee G, Irlitti A, Ens B, et al: On the shoulder of the giant: A multi-scale mixed reality collaboration with 360 video sharing and tangible interaction, in *Proceedings of the 2019 CHI Conference on Human Factors in Computing Systems*, pp. 1–17, 2019.

Regenbrecht H, Meng K, Reepen A, Beck S and Langlotz T: Mixed voxel reality: Presence and embodiment in low fidelity visually coherent mixed reality environments, in *Mixed and Augmented Reality (ISMAR) 2017 IEEE International Symposium on*, pp. 90–99, 2017.

Rhee T, Petikam L, Allen B and Chalmers A: Mr360, mixed reality rendering for 360 panoramic videos, in *IEEE Transactions on Visualization & Computer Graphics*, pp. 1379–1388, 2017.

Ruffaldi E and Brizzi F: CoCo—A framework for multicore visuohaptic in mixed reality, in *Salento AVR, 3rd International Conference on Augmented Reality, Virtual Reality and Computer Graphics*. New York: Springer, pp. 339–357, 2016.

Tang A, Biocca F and Lim L: Comparing differences in presence during social interaction in augmented reality versus virtual reality environments; an exploratory study, *Presence*: 204–208, 2004.

Tarko J, Tompkin J and Richardt C: Omnimr: Omnidirectional mixed reality with spatially-varying environment reflections from moving 360 video cameras, in *IEEE Conference on Virtual Reality and 3D User Interfaces*, pp. 1177–1178, 2019.

Teo T, Lawrence L, Lee G, Billinghurst M and Adcock M: Mixed reality remote collaboration combining 360 video and 3D reconstruction, in *Proceedings of the 2019 CHI Conference on Human Factors in Computing Systems*, pp. 1–14, 2019.

Velamkayala E, Zambrano M and Li H: Effects of hololens in collaboration: A case in navigation tasks, in *Proceedings of the Human Factors and Ergonomics Society Annual Meeting* 61: 2110–2114, 2017.

Xia T, Leonard S, Deguet A, Whitcomb L and Kazanzides P: Augmented reality environment with virtual fixtures for robotic telemanipulation in space, in *Proceedings of the 2012 IEEE/RSJ International Conference on Intelligent Robots*, pp. 5059–5064, 2012.

Touch, Data Gloves

Zhao Y, Ham J and van der Van der list J: Persuasive virtual touch: The effect of artificial social touch on shopping behavior in virtual reality, in *International Workshop on Symbiotic Interaction*, Springer, pp. 98–109, 2018.

Haptics, Kinesthetics

Antonakoglou K, Xu X, Steinbach E, Mahmoodi T and Dohler M: Toward haptic communications over the 5g tactile internet, *IEEE Communications Surveys and Tutorials* 20(4): 3034–3039, 2018.

Graziano A, Tripicchio P, Ruffaldi E and Avizzano C: A haptic datasuit for controlling humanoid robots, in *Proceedings of the 47th International Symposium on Robotics*, pp. 1–8, 2016.

Gwilliam J, Mahvash M, Vagvolgyi B, Vacharatet A, et al: Effects of haptic and graphical force feedback on teleoperated palpation, in *Proceedings of the 2009 IEEE International Conference on Robotics and Automation*, pp. 677–682, 2009.

Haddadin S and Croft E: Physical human–robot interaction, in *Springer Handbook of Robotics*, B. Siciliano and O. Khatib, Eds. Cham: Springer International Publishing, pp. 1835–1874, 2016.

Linkenauger S, Bulthoff H and Mohler B: Virtual arm's reach influences perceived distances but only after experience reaching, *Neuropsychologia* 70: 393–401, 2015.

Nitsch V and Farber B: A meta-analysis of the effects of haptic interfaces on task performance with teleoperation systems, *IEEE Transactions Haptics* 6(4): 387–398, October–December 2013.

Peppoloni L, Filippeschi A, Ruffaldi E and Avizzano C: A novel 7 degrees of freedom model for upper limb kinematic reconstruction based on wearable sensors, in *Proceedings of the IEEE 11th International Symposium on Intelligent Systems and Informatics*, pp. 105–110, 2013.

Sigrist R, Rauter G, Riener R and Wolf P: Augmented visual, auditory, haptic, and multimodal feedback in motor learning: A review, *Psychonomic Bulletin & Review* 20(1): 21–53, 2013.

Gesture Recognition, Position Tracking

Alem L and Li J: A study of gestures in a video-mediated collaborative assembly task, in *Advances in Human-Computer Interaction*, pp. 1, 2011.

Allison R, Eizenman M and Cheung B: Combined head and eye tracking system for dynamic testing of the vestibular system, *IEEE Transactions on Biomedical Engineering* 43(11): 1073–1082, 1996.

Amores J, Benavides X and Maes P: Showme, A remote collaboration system that supports immersive gestural communication, in *Proceedings of the 33rd Annual ACM Conference Extended Abstracts on Human Factors in Computing Systems*, pp. 1343–1348, 2015.

Huang W and Alem L: Supporting hand gestures in mobile remote collaboration: A usability evaluation, in *Proceedings of the 25th BCS Conference on Human-Computer Interaction*, pp. 211–216, 2011.

Kirk D and Stanton Fraser D: Comparing remote gesture technologies for supporting collaborative physical tasks, in *Proceedings of the SIGCHI Conference on Human Factors in Computing Systems*, pp. 1191–1200, 2006.

Renner R, Velichkovsky B, and Helmert J: The perception of egocentric distances in virtual environments—A review, *ACM Computing Surveys* 46(2): 1–40, 2013.

Tecchia F, Alem L and Huang W, 3D helping hands: A gesture based mr system for remote collaboration, in *Proceedings of the 11th ACM SIGGRAPH International Conference on Virtual-Reality Continuum and its Applications in Industry*, pp. 323–328, 2012.

Teo T, Lee G, Billinghurst M and Adcock M: Hand gestures and visual annotation in live 360 panorama-based mixed reality remote collaboration, in *Proceedings of the 30th Australian Conference on Computer-Human Interaction*, pp. 406–410, 2018.

Wood E, Taylor J, Fogarty J, Fitzgibbon A and Shotton J: Shadowhands: High-fidelity remote hand gesture visualization using a hand tracker, in *Proceedings of the 2016 ACM International Conference on Interactive Surfaces and Spaces*, pp. 77–84, 2016.

Yu J, Noh S, Jang Y, Park G and Woo W: A hand-based collaboration framework in egocentric coexistence reality, in *2015 12th International Conference on Ubiquitous Robots and Ambient Intelligence (URAI)*, pp. 545–548, 2015.

Master Controls

Hentout A, Benbouali R, Akli I, Bouzouia B and Melkou L: A telerobotic human/robot interface for mobile manipulators: A study of human operator performance, in *International Conference on Control, Decision and Information Technologies, CoDIT 2013*, pp. 641–646, 2013.

Hybl M: Design and development of a handheld robot controller with telepresence capabilities, in *26th Annual Conference on Student Electrical Engineering, Information Science and Communication Technologies Proceedings of the 26th Conference Student EEICT*, pp. 90–94, 2020.

Iwai D, Matsukage R, Aoyama S, Kikukawa T and Sato K: Geometrically consistent projection-based tabletop sharing for remote collaboration, *IEEE Access* 6: 6293–6302, 2017.

Keyes B, Micire M, Drury J and Yanco H: *Improving Human-Robot Interaction through Interface Evolution*, Edited by Daisuke Chugo, Kwansei Gakuin University, Japan, Intechopen.com, 2010.

Song W, Guo X, Jiang F, et al: Teleoperation humanoid robot control system based on kinect sensor, in *Proceedings of the 4th International Conference on Intelligent Human-Machine Systems and Cybernetics* 2: 264–267, 2012.

TELEOPERATION APPLICATIONS

Telesurgery

Kitagawa M, Dokko D, Okamura A and Yuh D: Effect of sensory substitution on suture-manipulation forces for robotic surgical systems, *The Journal of Thoracic and Cardiovascular Surgery* 129: 151–158, 2005.

Teleoperated Assembly

Aschenbrenner D, Leutert F, Çençen A, Verlinden J, et al: Comparing human factors for augmented reality supported single and cooperative repair operations of industrial robots, *Frontiers in Robotics and AI* 6: 37, 2019.

Bowman D and Hodges L: An evaluation of techniques for grabbing and manipulating remote objects in immersive virtual environments, *Symposium on Interactive 3D Graphics* 182, 1997.

Brizzi F, Peppoloni L, Ruffaldi E, et al: Effects of augmented reality on the performance of teleoperated industrial assembly tasks in a robotic embodiment, *IEEE Transactions On Human-Machine Systems* 48(2): 197–206, April 2018.

Peppoloni L, Brizzi F, Ruffaldi E and Avizzano C: Augmented reality-aided tele-presence system for robot manipulation in industrial manufacturing, in *Proceedings of the 21st ACM Symposium on Virtual Reality Software and Technology*, pp. 237–240, 2015.

Tang A, Owen C, Biocca F and Mou W: Comparative effectiveness of augmented reality in object assembly, in *Proceedings of the SIGCHI Conference on Human Factors in Computing Systems*, pp. 73–80, 2003.

Operations in Space and Undersea

Goecks V, Chamitoff G, Borissov S, Probe A, et al: Virtual reality for enhanced 3D astronaut situational awareness during robotic operations in space, in *AIAA Information Systems-AIAA Infotech at Aerospace*, January, 2017.

Keesey C: NASA eyes GPS at the moon for artemis missions, 2019. Available: https://www.nasa.gov/feature/goddard/2019/nasaeyes-gps-at-the-moon-for-artemis-missions.

McHenry C, Hunt T, Young W, Bhagavatula U, et al: Evaluation of pre-flight and on orbit training methods utilizing virtual reality, *AIAA SciTech*, published on line 5 Jan, 2020.

McHenry C, Spencer J and Chamitoff G: Predictive XR telepresence for robotic operations in space, in *IEEE Aerospace Conference (AeroConf)*, 2021.

Williams N: Risk of inadequate design of human and automation/robotic integration, 2018. Available: http://humanresearchroadmap.nasa.gov/risks.

Flight Simulation

Mendes D, Medeiros D, Sousa M, Ferreira R, et al: Mid-air modeling with boolean operations in VR, in *2017 IEEE Symposium on 3D User Interfaces (3DUI)*, pp. 154–157, 2017.

Driving Simulation

Mosiello G, Kiselev A and Loutfi A: Using augmented reality to improve usability of the user interface for driving a telepresence robot, *Journal of Behavioral Robotics* 4(3): 174–181, 2013.

Hazardous Environments

Zalud L, Kopecny L and Burian F: Orpheus reconnissance robots, in *IEEE International Workshop on Safety, Security and Rescue Robotics* [online]. IEEE, 31–34 [ref. 2020-03-12]. https://doi.org/10.1109/SSRR.2008.4745873. ISBN 9781424420315, 2008.

Zalud L, Kopecny L, Burian F and Florian A: Cassandra—heterogeneous reconnaissance robotic system for dangerous environments, in *IEEE/SICE International Symposium on System Integration (SII)*, pp. 1275–1280, 2011.

GROUP INTERACTION

2-Person Collaboration

Anton D, Kurillo G and Bajcsy R: Augmented telemedicine platform for real-time remote medical consultation, in *23rd International Conference on MultiMedia Modeling (MMM)*, pp. 77–89, 2017.

Gao L, Bai H, Lee G and Billinghurst M: An oriented point-cloud view for mr remote collaboration, in *SIGGRAPH ASIA 2016 Mobile Graphics and Interactive Applications*, pp. 8, 2016.

Kim S, Billinghurst M and Lee G: The effect of collaboration styles and view independence on video-mediated remote collaboration, *Computer Supported Cooperative Work (CSCW)* 27(3–6): 569–607, 2018.

Kolkmeier J, Harmsen E, Giesselink S, Reidsma D, et al: With a little help from a holographic friend: The openimpress mixed reality telepresence toolkit for remote collaboration systems, in *Proceedings of the 24th ACM Symposium on Virtual Reality Software and Technology VRST '18*, pp. 26:1–26:11, 2018.

Kunert A, Kulik A, Beck S and Froehlich B: Photoportals: Shared references in space and time, in *Proceedings of the 17th ACM Conference on Computer Supported Cooperative Work & Social Computing*, pp. 1388–1399, 2014.

Lee G, Teo T, Kim S and Billinghurst M: A user study on MR remote collaboration using live 360 video, in *IEEE International Symposium on Mixed and Augmented Reality (ISMAR)*, pp. 153–164, 2018.

Lee H, Ha G, Lee S and Kim S: A mixed reality telepresence platform to exchange emotion and sensory information based on mpeg-v standard, in *Virtual Reality (VR) 2017 IEEE*, pp. 349–350, 2017.

Sodhi R, Jones B, Forsyth D, Bailey B and Maciocci G: Bethere–3D mobile collaboration with spatial input, in *Proceedings of the SIGCHI Conference on Human Factors in Computing Systems*, pp. 179–188, 2013.

Wolff R, Roberts D, Steed A and Otto O: A review of telecollaboration technologies with respect to closely coupled collaboration, *International Journal of Computer Applications in Technology* 29(1): 11–26, 2007.

Multi-Person Teleconferencing

Anton, D, Kurillo G and Bajcsy R: User experience and interaction performance in 2D/3D telecollaboration, in *Future Generation Computer Systems-The International Journal of Escience* 82, pp. 77–88, May 2018.

Beck S, Kunert A, Kulik A and Froehlich B: Immersive group-to-group telepresence, *IEEE Transactions on Visualization and Computer Graphics* 19(4): 616–625, April 2013.

Dijkstra-Soudarissanane S, El Assal K, et al: Multi-sensor capture and network processing for virtual reality conferencing, in *10th ACM Multimedia Systems Conference (ACM MMSys)*, pp. 316–319, 2019.

Gerhard M, Moore D and Hobbs D: Embodiment and copresence in collaborative interfaces, *International Journal of Human-Computer Studies* 61(4): 453–480, 2004.

Ibayashi H, Sugiura Y, Sakamoto D, Miyata N, et al: Dollhouse VR, a multi-view multi-user collaborative design workspace with vr technology, in *SIGGRAPH Asia 2015 Emerging Technologies*, pp. 1–2, 2015.

Kasahara S, Ando M, Suganuma K and Rekimoto J: Parallel eyes: Exploring human capability and behaviors with paralleled first-person view sharing, in *Proceedings of the 2016 CHI Conference on Human Factors in Computing Systems*, pp. 1561–1572, 2016.

Lee Y and Yoo B: XR collaboration beyond virtual reality, work in the real world, *Journal of Computational Design And Engineering* 8(2): 756–772, April 2021.

Sra M, Mottelson A and Maes P: Your place and mine: Designing a shared VR experience for remotely located users, in *Proceedings of the 2018 on Designing Interactive Systems Conference 2018*, pp. 85–97, 2018.

Wienrich C, Schindler K, Döllinqer N, Kock S and Traupe O: Social presence and cooperation in large-scale multi-user virtual reality-the relevance of social interdependence for location-based environments, in *2018 IEEE Conference on Virtual Reality and 3D User Interfaces (VR)*, pp. 207–214, 2018.

Learning

Anton D, Kurillo G and Bajcsy R: User experience and interaction performance in 2D/3D telecollaboration, *Future Generation Computer Systems* 82: 77–88, 2018.

Bessa M, Melo M, Narciso D, Barbosa L and Vasconcelos-Raposo J: Does 3D 360 video enhance user's VR experience?: An evaluation study, in *Proceedings of the XVII International Conference on Human Computer Interaction*, pp. 16, 2016.

Billinghurst M, Nassani A and Reichherzer C: Social panoramas: using wearable computers to share experiences, in *SIGGRAPH Asia 2014 Mobile Graphics and Interactive Applications*, pp. 25, 2014.

Morozov M: Augmented reality in military, AR can enhance warfare and training, 2018. [Online]. Available: https://jasoren.com/augmentedreality-military.

Shmuelof L, Krakauer J and Mazzoni P: How is a motor skill learned? Change and invariance at the levels of task success and trajectory control, *Journal of Neurophysiology* 108(2): 578–594, 2012.

Social Behavior

Bailenson J, Blascovich J, Beall A and Loomis J: Interpersonal distances in virtual environments, *Personality and Social Psychology Bulletin* 29(7): 819–833, 2003.

Chun K and Campbell J: Dimensionality of the rotter interpersonal trust scale, *Psychological Reports* 35(3): 1059–1070, 1974.

Latoschik M, Kern F, Stauffert J, Bartl A, et al: Not alone here?! Scalability and user experience of embodied ambient crowds in distributed social virtual reality, *IEEE Transactions on Visualization and Computer Graphics* 25(5): 2134–2144, 2019.

Ranieri N, Bazin J, Martin T, Laffont P, et al: An immersive bidirectional system for life-size 3D communication, in *Proceedings of the 29th International Conference on Computer Animation and Social Agents*, pp. 89–96, 2016.

Roth D, Klelnbeck C, Feigl T, Mutschler C and Latoschik M: Beyond replication: Augmenting social behaviors in multi-user virtual realities, in *2018 IEEE Conference on Virtual Reality and 3D User Interfaces (VR)*, pp. 215–222, 2018.

Teo T, Lee G, Billinghurst M and Adcock M: Investigating the use of different visual cues to improve social presence within a 360 mixed reality remote collaboration, in *The 17th International Conference on Virtual-Reality Continuum and its Applications in Industry*, pp. 1–9, 2019.

Zibrek K, Kokkinara E and Mcdonnell R: The effect of realistic appearance of virtual characters in immersive environments—does the character's personality play a role? *IEEE Transactions on Visualization and Computer Graphics* 24(4): 1681–1690, April 2018.

Theater, Gaming

Leap M: *Welcome to the Shared Reality Magic—Leap*, [online]. Available: https://creator.magicleap.com/learn/guides/leap-welcome-to-shared.

ASSISTING PEOPLE

Physically Disabled

Klinger E, Cherni H and Joseph P: Impact of contextual additional stimuli on the performance in a virtual activity of daily living (vADL) among patients with brain injury and controls, *International Journal on Disability and Human Development*, Veroffenlicht von De Gruyter, 21 July, 2014.

Social Robotics

Alhaddad AY, Cabibihan JJ and Bonarini A: Head impact severity measures for small social robots thrown during meltdown in autism, *International Journal of Social Robotics* 11(2): 255–270, 2019.

Block A and Kuchenbecker K: Softness, warmth, and responsiveness improve robot hugs, *International Journal of Social Robotics* 11(1): 49–64, 2019.

Brohl C, Nelles J, Brandt C, Mertens A and Nitsch V: Human-robot collaboration acceptance model: Development and comparison for Germany, Japan, China and the USA, *International Journal of Social Robotics*, 2019. https://doi.org/10.1007/s12369-019-00593-0.

Bruno B, Recchiuto C, Papadopoulos I, Saffiotti, A, Koulouglioti C, et al: Knowledge representation for culturally competent personal robots: Requirements, design principles, implementation, and assessment, *International Journal of Social Robotics* 11(3): 515–538, 2019.

Carlson Z, Lemmon L, Higgins M, Frank D and Shahrezaie R: Perceived mistreatment and emotional capability following aggressive treatment of robots and computers, *International Journal of Social Robotics*, October 24: 1–13, 2019. [Online]. Available: https://doi.org/10.1007/s12369-019-00599-8.

Coghlan S, Vetere F, Waycott J and Neves B: Can social robots make us kinder or crueler to humans and animals? *International Journal of Social Robotics*, https://doi.org/10.1007/s12369-019-00599-8.2019.

Erlich S and Cheng G: A feasibility study for validating robot actions using EEG-based error-related potentials, *International Journal of Social Robotics* 11(2): 271–283, 2019.

Fattal C, Leynaert V, Laffont I, Baillet A, Enjalbert M and Leroux C: SAM, an assistive robotic device dedicated to helping persons with quadriplegia: Usability study, *International Journal of Social Robotics* 11(1): 89–103, 2019.

Hamandi M, Hatay E and Fazli P: Predicting the target in human-robot manipulation tasks, in *2018 Proceedings of the International Conference on Social Robotics*, pp. 580–587, 2018.

Heimerdinger M and LaViers A: Modeling the interactions of context and style on affect in motion perception: Stylized gaits across multiple environmental contexts, *International Journal of Social Robotics* 11(3): 495–513, 2019.

Homma Y and Suzuki K: A robotic brush with surface tracing motion applied to the face, in *Proceedings of the International Conference on Social Robotics*; See also *Lecture Notes in Computer Science* Vol 11357, 2018.

Hoorn J, Konijn E, and Pontier M: Dating a synthetic character is like dating a man, *International Journal of Social Robotics* 11(2): 235–253, 2019.

Ismail L, Verhoeven T, Dambre J and Wyffels F: Leveraging robotics research for children with autism: A review, *International Journal of Social Robotics* 11(3): 389–410, 2019.

Johnson D and Cuijpers R: Investigating the effect of a humanoid robot's head position on imitating human emotions, *International Journal of Social Robotics* 11(1): 65–74, 2019.

Jonaiti M and Henaff P: Robot-based motor rehabilitation in autism: A systematic review, in *Proceedings of the International Conference on Social Robotics*, pp. 1–12, 2018.

Karatas N, Tamura S, Fushiki M and Okada M: The effects of driving agent gaze following behaviors on human autonomous car interaction, in *Proceedings of the International Conference on Social Robotics*, pp. 541–550, 2018.

Karunarathne D, Morales Y, Nomura T, Kanda T, Ishiguro H and Karunarathne D: Will older adults accept a humanoid robot as a walking partner, *International Journal of Social Robotics* 11(2): 343–348, 2019.

Kaushik R and LaViers A: Imitating human movement using a measure of verticality to animate low degree-of-freedom non-humanoid virtual characters, in *2018 Proceedings of the International Conference on Social Robotics*, pp. 599–598, 2018.

Kaushik R and LaViers A: Imitation of human motion by low degree-of-freedom simulated robots and human preference for mappings driven by spinal, arm and leg activity, *International Journal of Social Robotics* 11: 765–782, 2019.

Komatsubara T, Shiomi M, Kaczmarek T, Kanda T and Ishiguro H: Estimating children's social status through their interaction activities in classrooms with a social robot, *International Journal of Social Robotics* 11(1): 35–48, 2019.

Korn O (editor): *Social Robots: Technological, Societal and Ethical Aspects of Human-Robot Interaction* (Human–Computer Interaction Series). Springer 2019.

Kostavelis I, Vasileiadis E, Skartados E, Kargakos A, et al: Understanding of human behavior with a robotic agent through daily activity analysis, *International Journal of Social Robotics* 11(3): 437–462, 2019.

Li H, Yihun Y and He H: Magic hand: in-hand perception of object characteristics for dexterous manipulation, in *Proceedings of the International Conference on Social Robotics*, pp. 523–532, 2018.

Liu T, Wang J, Hutchinson S and Meng M: Skeleton-based human action recognition by pose specificity and weighted voting, *International Journal of Social Robotics* 11(2): 219–234, 2019.

Lupowski P, Rybka M, Dziedic D and Wlodarczyk W: The background context condition for the uncanny valley hypothesis, *International Journal of Social Robotics* 11(1): 25–33, 2019.

Martins G, Santos L and Dias J: User-adaptive interaction in social robots: A survey focusing on non-physical interaction, *International Journal of Social Robotics* 11(1): 185–205, 2019.

Moro C, Lin S, Nejat G and Mihailidis A: Social robots and seniors: A comparative study on the influence of dynamic social features on human-robot interaction, *International Journal of Social Robotics* 11(1): 5–24, 2019.

Palanica A, Thommandram A and Fossat Y: Adult verbal comprehension performance is better from human speakers than social robots, but only for easy questions, *International Journal of Social Robotics* 11(2): 359–369, 2019.

Papenmeier F, Uhrig M and Kirsch A: Human understanding of robot motion: The role of velocity and orientation, *International Journal of Social Robotics* 11(1): 75–88, 2019.

Parviainen J, Turja T and VanAerschot L: Robots and human touch in care: Desirable and non-desirable robot assistance, in *Proceedings of the International Conference on Social Robotics*, pp. 533–540, 2019.

Portugal D, Alvito P, Christodoulou E, Samaras G and Dias J: A study on the deployment of a service robot in an elderly care center, *International Journal of Social Robotics* 11(2): 317–341, 2019.

Radmard S, Moon A and Croft E: Impacts of visual occlusion and its resolution in robot-mediated social collaborations of social robots, *International Journal of Social Robotics* 11(1): 105–121, 2019.

Ruijten P, Haans A, Ham J and Midden C: Perceived human-likeness of social robots: Testing the Rasch model as a method for measuring anthropomorphism, *International Journal of Social Robotics* 11(4): 477–494, 2019.

Shariati A, Shahab M, Meghdari A, Nobaveh AA, et al: Virtual reality social robot platform: A case study on Arash social robot, in *Proceedings of the International Conference on Social Robotics*, pp. 551–560, 2018.

Spatola N, Belletier C, Chausse P, Augustinova M, Normand A, Barra V, Ferrand L and Huguet P: Improved cognitive control in presence of anthropomorphized robots, *International Journal of Social Robotics* 11(3): 463–476, 2019.

Sprute D, Tonnies K and Konig M: A study on different user interfaces for teaching virtual borders to mobile robots, *International Journal of Social Robotics* 11(3): 373–388, 2019.

Stroessner S and Benitez J: The social perception of humanoid and non-humanaoid robots: Effects of gendered and machinelike features, *International Journal of Social Robotics* 11(2): 305–315, 2019.

Wang B and Rau P-L: Influence of embodiment and substrate of social robots on users' decision-making and attitude, *International Journal of Social Robotics* 11(3): 411–421, 2019.

Wang L, Du Z, Dong W, Shen Y and Zhao G: Hierarchical human machine interaction learning for a lower extremity augmentation device, *International Journal of Social Robotics* 11(1): 123–139, 2019.

Willemse C and van Erp J: Social touch in human–robot interaction: Robot-initiated touches can induce positive responses without extensive prior bonding, *International Journal of Social Robotics* 11(2): 285–304, 2019.

Wollherr D and Turnwald A: Human-like motion planning based on game theoretic decision making, *International Journal of Social Robotics* 11(1): 151–170, 2019.

Yamashita Y, Ishihara H, Ikeda T and Asada M: Investigation of causal relationship between touch sensations of robots and personality impressions by path analysis, *International Journal of Social Robotics* 11(1): 141–150, 2019.

Yoon HS, Jang J and Kim J: Multi-pose face recognition method for social robot, in *Proceedings of the International Conference on Social Robotics*, pp. 609–619, 2018.

Yoshikawa Y, Kumazake H, Matsumoto Y, Miyao M, Ishiguru H and Shimaya J: Communication support vis a tele-operated robot for easier talking: Case/laboratory study of individuals with/without autism spectrum disorder. *International Journal of Social Robotics* 11(1): 171–184, 2019.

Zhang J, Zhang H, Dong C, Hang F, Liu Q and Song A: Architecture and design of a wearable robotic system for body posture monitoring, correction and rehabilitation assist, *International Journal of Social Robotics* 11(3): 423–436, 2019.

COMMERCIAL APPLICATIONS

SHOPPING

Ploydanaim K, van den Puttelaar J, et al: Using a virtual store as a research tool to investigate consumer in-store behavior, *Journal of Visualized Experiments*, 125, July 24, 2017.

Sikström E, Høeg E, Mangano L, Nilsson N, et al: Shop'til you hear it drop: Influence of interactive auditory feedback in a virtual reality supermarket, Aalborg Universitet, in *Proceedings of the ACM*, Conf on Virtual Relaity Software, 355=356, November 2016.

Xi N and Hamari J: Shopping in virtual reality – A literature review and future agenda, *Journal of Business Research* 134: 37–58, Sept 2021.

Xi N and Hamari J: Shopping in virtual reality, a literature review and future agenda, *Journal of Business Research* 134: 37–58, September 2021.

Marketing and Media

Global Social Robot Market 2018–2023: Product Innovations and new launches will intensify competitiveness. www.prnewswire.com/news-releases/global-social.

Kishore S, Navarro X, Dominguez E, De La Peña N and Slater M: Beaming into the news: A system for and case study of tele-immersive journalism, in *IEEE Computer Graphics and Applications* 10, 2016.

Li R: Big packet protocol: Advances the internet with in-network services and functions. *MMTC Communications–Frontiers*, mmc.committees.comsoc.org, 2019.

Wedel M, Bigne E and Zhang J: Virtual and augmented reality–Advancing research in consumer marketing, *International Journal of Research in Marketing* 37(3): 443–465, 2020.

Tourism

Loureiro S, Guerreiro J and Ali F: 20 years of research on virtual reality and augmented reality in tourism context: A text-mining approach, *Tourism Management* 77, April 2020.

Wei W: Research progress on virtual reality (VR) and augmented reality (AR) in tourism and hospitality a critical review of publications from 2000 to 2018, *Journal of Hospitality and Tourism Technology* 10(4): 539–570, 2019.

LEGAL AND PHILOSOPHICAL ISSUES

Legal

Botrugno C: Telemedicine in daily practice, addressing legal challenges while waiting for an EU regulatory framework, *Health Policy and Technology* 7(2): 131–136, 2018.

European Commission: Commission staff working document on the applicability of the existing EU legal framework to telemedicine services, in *Innovative Healthcare for the 21st Century 2012*, June 6, 2012.

European Commission: *Market study on telemedicine*, European Commission, 2018.

Jang S, Lee K, Hong Y, Kim J and Kim S: Economic evaluation of robot-based telemedicine consultation services, *Telemedicine and e-Health* 26(9): 1134–1140, 2020.

Lin C: Mobile telemedicine: A survey study, *Journal of Medical Systems* 36(2): 511–520, 2012.

Raposo V: Telemedicine: The legal framework (or the lack of it) in Europe, *GMS Health Technology Assessment* 12, Aug 16, 2016.

Reindl A, Rudigkeit N, Ebers M, Tröbinger M, et al: Legal and technical considerations on unified, safe and data-protected haptic telepresence in healthcare, in *2021 IEEE International Conference on Intelligence and Safety for Robotics (ISR)*, pp. 239–243, 2021.

Philosophical

Chalmers D: *Reality Plus*, New York, W. W. Norton & Company, 2022.

Index

M

N

O

P

Q

R

S